THE SEARCH FOR THE PAST

L.B. HALSTEAD

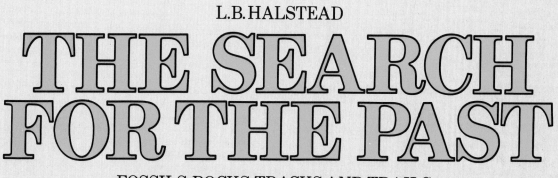

THE SEARCH FOR THE PAST

FOSSILS, ROCKS, TRACKS AND TRAILS
THE SEARCH FOR THE ORIGIN OF LIFE

Doubleday & Company Inc.
Garden City, New York
1982

Doubleday & Company Inc.
Garden City, New York

Editors: Louisa McDonnell,
Janet Sacks, Jana Gough
Art Direction: David Pearce
Design: Jane Storrar
Picture Research: Caroline Lucas
Glossary: J. G. Collins
Index: Valerie Lewis Chandler
Typesetting: Input Typesetting Limited
Reproduction: F.E. Burman Limited

Made by Roxby Prehistory Limited
Roxby & Lindsey Press
98 Clapham Common Northside,
London SW4 9SG

Library of Congress Cataloging in
Publication Data

Halstead, L.B.
The Search for the Past

Bibliography: P.
Includes index
1. Paleontology. 1. Title
QE711.2.H34   1982   560   82-45203
ISBN 0-385-18212-0   AACR2

First American Edition 1982

ACKNOWLEDGEMENTS

Heartfelt thanks are due to my two friends, Professor John R.L. Allen FRS and Professor Zofia Kielan-Jaworowska; my colleagues, Mr John Collins, Drs Roland Goldring, Bruce Sellwood and Peter Worsley helped greatly in various ways, and Mrs Irene Gillett, Mrs Elizabeth Wyeth, Mrs Angela Broughton and Mrs Gillian Coward performed the Trojan service of typing the manuscript. *L.B. Halstead*
The author and publishers would also like to thank the following people for illustrations reproduced in the book.

Photographs
Academy of Applied Science, Boston/Syndication International 198 (bottom & top right); Aerofilms 28 (left), 29 (left); Alberta Map and Air Photo Distribution Centre 107; Prof. J. R. L. Allen 106, 115 (bottom left), 116 (top), 117; American Museum of Natural History 47 (right), 78 (bottom), 99 (top), 133 (bottom); American Philosophical Society 72 (bottom), 79 (bottom); Animal Photography Ltd 155 (bottom); Ardea 138 (top), 197 (bottom); Dr Elso Barghoorn, Harvard University 87 (right); British Library 74-5; British Museum (Natural History) 48-9, 70-1, 74-5, 77, 85, 141 (centre), 164-5, 177 (top), 180-1, 199 (right); Frank M. Carpenter, Museum of Zoology, Cambridge, Mass. 37; J. Allan Cash 16 (bottom left); John Cleare/Mountain Camera 28 (right); Bruce Coleman Ltd 19 (top), 40 (centre), 41, 88-9 (top), 129, 143 (top), 143 (bottom); Committee for Aerial Photography, Cambridge 34 (bottom right); Dr S. Conway Morris (top, bottom left & right); Dr Laurence Cook, University of Manchester 80 (left & right); Crown Copyright 179; P. W. Currie 133 (top); Henry de Lumley (J. Vertut) 168 (top); Paul R. Devall 58 (bottom left & right), 59 (bottom); D. Dickins 14 (top & bottom left); Earl Douglass Papers, Special Collections Dept., University of Utah 47 (left); Ecology Pictures 16 (right), 23 (top & bottom), 24 (bottom left & bottom centre), 32 (left), 33 (top right), 49 (top right); Eden Films 199 (left); Dr Diane Edwards 119 (bottom); Mark Ellen 29 (right); E. M. Friis, Bedford College, London University 119 (top); Geodetic Institute 34 (left); Geological Society of London 176 (bottom), 177 (centre right); Geological Survey, U.S. Dept. of Interior 78 (top); Dr R. Goldring 48 (bottom left), 50 (top right), 53 (top left & right), 89 (bottom), 92-3; Tim R. Good 35 (top centre); L. B. Halstead 8 (D. Pearce), 14 (bottom right), 20 (top, D. Pearce), 21 (bottom centre, D. Pearce), 21 (bottom right), 22 (top & bottom right), 24 (bottom right), 26-7 (bottom), 35 (top right), 40 (left), 44 (left), 52 (top left), 54 (top & bottom), 55, 58 (top), 66, 67 (top & bottom), 111 (top, bottom left & right), 112, 113, 130, 135, 141 (top right), 144, 194, 196 (bottom); Illustrated London News 177 (centre left & bottom); Imitor Ltd 45, 103 (top, left & right), 104, 105, 108, 109, 111 (bottom right), 122 (top & bottom right), 141 (bottom right), 170-1 (top); Institute of Geological Sciences, Edinburgh 101 (left); reproduced by permission of the Director, Institute of Geological Sciences, London: Crown copyright reserved 12, 13 (top), 13 (bottom left), 192; Peter Kain/© Richard Leakey 22 (left), 156, 158, 159, 168 (bottom left), 169, 171 (bottom left), 172 (centre), 173 (top); Prof. Zofia Kielan-Jaworowska 60, 61, 62, 63, 64; Dick Kenny 40 (right); Prof. Dr Bernard Krebs, Berlin 145; Prof. Dr Emil Kuhn-Schnyder 140 (bottom); Pat Morris 36; Marion & Tony Morrison 191 (top); Museum beim Solenhofer, Maxberg 50 (top left); Museum für Naturkunde, Berlin 83 (top); NASA 178; Natural History Photographic Agency 141 (top left); National Museums of Kenya 43 (bottom right); National Museum of Wales 69; Novosti Press 43 (top), 163 (bottom); Dr Kenneth Oakley 68 (left), 68 (right © County Museum, Liverpool); C. Parsons 96-7; Peale Museum, Baltimore 72-3; David Pearce 17, 19 (bottom left), 48 (top left), 52 (top right), 53 (top left); Penn Monthly/Sterling Library, Yale, 79 (top); Punch Publications 82 (top); John Reader/National Geographical Society/Olduvai Research Fund 157; RIDA Picture Library 10 (bottom right), 16 (top left), 24 (top), 84, 87 (right), 88 (bottom), 102, 132 (right), 171 (right); Gordon Riley/Prof. A. J. Cain 81; Royal Dental Hospital 98 (bottom); Science Museum, London 182-3; Dr Andrew Scott 122 (top & bottom left, top & bottom centre); Dr Colin Scrutton 59 (top); Seaphot/Planet Earth Pictures 155 (top), 188, 196 (top), 197 (top); Dr Bruce Sellwood 29 (centre), 44 (top right), 115 (top & bottom left); Prof. A. J. Smith, Bedford College, London 181 (top centre); Smithsonian Institution 148, 149, 152, 153, 164, 165; Staatliches Museum für Naturkunde, Stuttgart 46; Dr A. D. Stewart 52 (bottom), 53 (bottom left); Dr C. B. Stringer 163 (top); Prof. Wilhelm Stürmer 42; Prof. L. P. Tatarinov 138 (bottom left & right), 139 (bottom left & right); Tubingen University 128; Dr V. Turek, Prague 100-01; Dr G. van Einden 193; J. Vertut 168 (top Collection Begonen), 172 (bottom & top – Musée d'Aquitaine), 173 (bottom); Dr Mary Wade, Queensland Museum, Brisbane 132 (left); J. L. Watkins, Reading University 38, 39, 50 (bottom), 176 (top); Wayne State University 43 (bottom left); Mary Weech 191 (bottom); Weidenfeld and Nicolson 76; Prof. H. B. Whittington 94 (top & bottom), 95 (top left); Dr P. Wilson 26 (top & bottom), 26-7 (top), 27 (bottom); Woods Hole Oceanographic Institution 187; Dr Peter Worsley 9, 18, 19 (bottom right), 21 (centre), 25, 32 (right), 33 (top & bottom left, bottom centre & bottom right), 34 (top right), 35 (left & bottom right), 44 (bottom right), 49 (bottom right), 115 (top right), 116 (bottom).

Artwork
Graham Allen 56, 57, 96-7, 100-01, 168-9; Eddi Gornall 104-5; Jenny Halstead 99, 110 (right); Ted Hammond 11, 30, 31, 62, 66-67, 80, 86, 133, 184; Aziz Khan 12-13, 16, 17, 54, 55; Edwina Keene 25, 29, 38, 40, 44, 53, 64, 68, 76-7, 82, 83, 90-1, 106, 108-9, 114, 116, 117, 120-1, 126-7, 130, 131, 134, 135, 136, 137, 139, 145, 146, 147, 154, 189 (bottom); Simon Roulstone 183, 184 (right), 185, 189 (top); Javier Sànchez 160, 161; Philip Wong 158, 159, 170; John Woodcock 15, 18, 23, 26, 42, 46-7, 50, 51, 94, 98, 103, 110, 111, 118, 119, 123, 124-5, 128, 129, 141, 142, 150, 151, 161, 162 (top right), 166 (top), 167 (top), 174-5, 180, 181, 186, 187, 189, 190, 191, 198, 199, 200 (top & centre). After the following authors: R. Mc N. Alexander 133; J. R. L. Allen 11, 23, 106, 114 (top left, right), 115, 116 (bottom), 117; J. E. Cronin *et al.* 162; R. Dart 158-9; A. J. de Ricqles 134 (left); R. Goldring 114 (bottom left); L. B. Halstead 98, 142, 150-1; P. H. Heckel & B. J. Witzke 116 (top); J. A. Hopson 134-5; J. Imbrie & K. P. Imbrie 190; Institute of Geological Sciences, London 190; A. F. King 50; L. Margulis 90-91; A. Milner 126-7; D. Norman 76-7, 199; R. Nursall 200 (centre); A. S. Romer 144; J. W. Schopf 86; A. Seilacher 51; R. Sloan & L. van Valen 146-7; J. R. Spotilla *et al.* 137 (top); A. J. Sutcliffe 54-5; A. D. Walker 42; E. K. Walton 123; P. E. Wheeler 137 (bottom); H. B. Whittington & S. Conway Morris 96-7. Every effort has been made to trace copyright-holders. It is hoped that any omission will be excused.

Chapter opening photographs
8 Ichthyosaur from Lyme Regis, Dorset, England, showing preservation of part of the intestine. 9 Palaeontologists await a helicopter to fly out dinosaur remains from Dinosaur Provincial Park, Alberta, Canada. 36 Rock face at Dinosaur National Monument, Utah, USA. 37 Insect larva preserved in Baltic amber. 60 Loading crated dinosaur remains onto a truck during a Polish-Mongolian expedition in the Gobi Desert. 61 Excavating a *Tarbosaurus* skeleton in the Gobi Desert. 84 Banded ironstone. 85 The crinoid *Pentacrinus*. 156 *Ramapithecus* and *Sivapithecus* remains from the Siwalik Hills, Pakistan. 157 The earliest evidence of a human family. 3.6 million year-old footprints at Laetoli, Tanzania. 178 Satellite photograph of Red Sea. 179 Photograph of dune-bedding at Mauchline Quarries, Scotland.

CONTENTS

CONTENTS

CHAPTER 1

READING THE ROCKS

The landscape of the Earth appears timeless
and unchanging. Yet on close examination of the rocks,
seemingly the most durable features of all,
another impression emerges. For the rocks show us that
the landscape is continually moving and altering:
sediments laid down in the seas now form the tops
of mountains. In the rocks are fossils, the remains of
plants and creatures which once inhabited the Earth,
revealing that life has evolved over hundreds of
millions of years. Every piece of rock, every fossil
fragment, is a tangible record of the history of the Earth.
Once the record of the rocks can be read, we can trace
the origin and evolution of life on Earth.

HOW TO READ THE ROCKS

It was not until the eighteenth century that geologists realized that the landscape was shaped by a continual cycle of erosion, deposition and uplift. Contemporary scientists and churchmen viewed this as heresy.

Rocks appear to be the most permanent features of the landscape, yet three-quarters of the land surface of the Earth is made up of sedimentary rocks. These, as their name implies, are made up of sediments which themselves are derived from the breakdown of pre-existing rocks.

For centuries nobody seriously thought about how rocks were formed – it was simply accepted that they were there and had been so since the creation of the world. One theory was that the rock strata with their fossilized remains had settled after the Biblical Flood had subsided. This Diluvian idea first appeared in early Christian times and persisted until well into the eighteenth century. In fact some books are still being published advocating a similar explanation of the geological record in terms of such global catastrophes.

However, occasionally a thinker would express doubts: Leonardo da Vinci, for example, realized that the presence of seashells in the Tuscany hills of central Italy meant that such places had once been at the bottom of the sea and had subsequently been raised up.

The first detailed attempt to understand the meaning of rocks was made by the Scotsman James Hutton (1726–97), after thirty years of observation and study of the rocks in Scotland and elsewhere. In 1785 he presented his conclusions to the Royal Society of Edinburgh. He recognized that sedimentary rocks were the result of erosion by the weathering of the land surface followed by the transport of materials by rivers and streams to the seas, where they were deposited. Thereafter the pressure and heat that resulted from their burial at depth led to the sediment being formed into solid

rock. Subsequent volcanic activity lifted up the rock to form dry land and the cycle began again.

Hutton emphasized that rocks could be explained in terms of processes that could be observed. This interpretation led him to conclude that the landscape was formed by these processes occurring in a never-ending cycle of erosion, deposition, uplift and erosion. Many of the leading scientists of his day attacked Hutton with vehemence, accusing him of trying to discredit the Scriptures. These attacks resulted in Hutton publishing his two-volume work, *Theory of the Earth, with Proofs and Illustrations*, in 1795. He concluded his monumental study with the words, 'We find no vestige of a beginning – no prospect of an end.' It was this kind of statement that incensed his critics because he had deliberately avoided any reference to

▲ Siccar Point, Scotland is a classic 'unconformity'. The Old Red Sandstone strata rest on older vertical Silurian rocks. James

Hutton realized that the Silurian rocks had been uplifted at an angle, eroded and the overlying strata subsequently deposited.

▲ The bands on the columns of the Roman temple of Jupiter Serapis, Italy were bored by marine shellfish.

▲ A granite vein cuts through a rock at Cataluña, Spain. Hutton saw from evidence like this that granite was once molten.

the origin of the world (and perhaps its end) at the hand of the Creator.

It was unfortunate that Hutton's writing was difficult to follow and it was not until 1802, when John Playfair (1748–1819), another Scotsman, published a more lucid account called *Illustrations of the Huttonian Theory of the Earth*, that the ideas really began to take hold. Playfair was a brilliant advocate and his arguments helped to make Hutton's views acceptable to both the scientific and ecclesiastical communities.

Final acceptance of the approach that Hutton advocated, however, came with Charles Lyell's (1797–1875) publication in 1830 of *Principles of Geology, Being an Attempt to Explain the Former Changes of the Earth's Surface, by Reference to Causes now in Operation*. It was Lyell who first said, 'The present is the key to the past.' This principle, which is known as 'uniformitarianism' because it stresses the uniformity of natural forces in past and present working in a process of slow, unending change, still remains the basic approach of all students of geology at the present day.

Among the evidence of an unending cycle of erosion, deposition and uplift, Hutton pointed to unconformities (breaks in the continuity of the rock strata) such as Siccar Point, Berwickshire, Scotland, where one set of strata was seen resting on an underlying set that was at a marked angle.

Hutton's further insistence that granite had once been molten, a view vigorously denied by his contemporaries, was demonstrated from his observation of veins of granite cutting through layers of sediment. The molten granite had baked the sediment, which appeared lighter where the granite had come in contact.

Charles Lyell was able to convince his readership of all this without challenging their religious beliefs. The Church even agreed to his appointment as Professor of Geology at the Church of England's institution, King's College, London. He recorded rates of erosion that could be observed taking place in his lifetime.

Among the striking evidences of change to which he was able to draw attention was the Roman temple of Jupiter Serapis, south of Naples, where there was a zone of columns that had been attacked by marine rock-boring bivalve shells. This could only have come about if, after the temple had been built, the sea had risen (or the land subsided) so that the columns became vulnerable to the attentions of these shellfish. Thereafter the sea retreated. The changes in sea-level over a mere 2,000 years provided incontrovertible evidence of the reality of the processes invoked by Hutton and consolidated by Lyell.

Charles Darwin (1809–82) calculated that a 500 foot (152 m) cliff would be eroded at the rate of 1 in (2.5 cm) per century and therefore it had taken 300 million years to erode the rocks overlying the early Cretaceous rocks of the Weald in south-east England. Darwin was quite wrong in this, his estimate being three times too large; however, it indicates the time span envisaged by the early geologists.

As Hutton was the first to point out, most types of sedimentary rocks can be understood in terms of the processes that are taking place at the present time, but it has nevertheless to be realized that although the basic principles may be applicable, they cannot be applied in every detail.

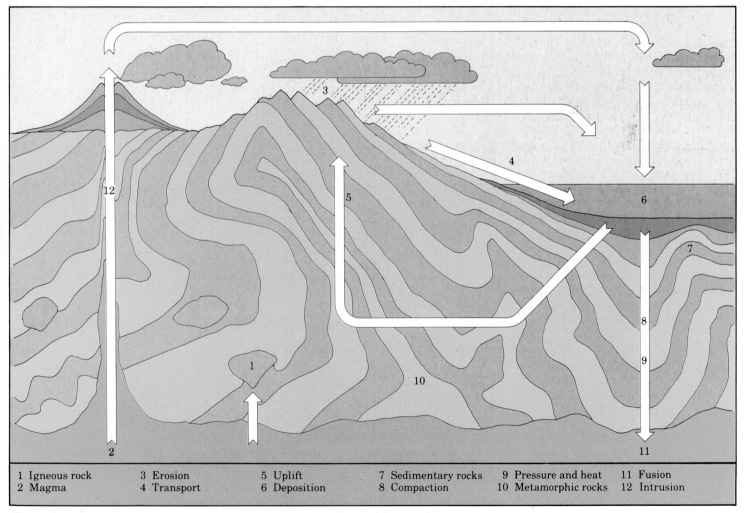

1 Igneous rock	3 Erosion	5 Uplift	7 Sedimentary rocks	9 Pressure and heat	11 Fusion
2 Magma	4 Transport	6 Deposition	8 Compaction	10 Metamorphic rocks	12 Intrusion

The rock cycle
Hutton and Lyell recognized a seemingly never-ending rock cycle. Erosion wears down the land and sediments are transported by wind and water and deposited. Pressure and heat at great depths cause mountain-building activity and new land is raised up.

THE FIRST ROCKS

The first rocks on Earth were igneous rocks, formed from molten material. Volcanic activity produces similar rocks today. All other types of rock ultimately derive from igneous rocks.

Three-quarters of all rocks on the land have been derived from pre-existing rocks in a continuous cyclical process (see pp 10–11). Ultimately these sediments can be seen to have originated from rocks that crystallized from a molten state. Rocks formed by this process, which involves an enormous amount of heat, are known as igneous rocks (from the Latin *ignis*, fire), and these were the first rocks to form on Earth.

The oldest known rocks on the Earth have come from space, from the asteroid belt situated between the orbits of Mars and Jupiter. These meteorites or 'falling stars' are 4,600 million years old; crystalline igneous rocks of the same age have been brought back from the Moon. From this evidence it seems reasonable to assume that the different bodies of our Solar System are all of the same age, hence it is thought that the Earth is also 4,600 million years old and was originally composed of igneous rocks.

The oldest known rock formed on the Earth is a pebble of a volcanic ash in a conglomerate (or pebble bed) from Isua, in West Greenland, and is 3,824 million years old. Evidently 800 million years of recycling have gone unrecorded.

The processes which produce igneous rocks are still in operation and can be observed wherever there are active volcanoes. When 'magma', a mixture of molten rock and hot gases, breaks through the Earth's crust, the gases escape and lavas pour over the surface. But if the magma reaches the surface very suddenly, the cooling will be so fast that the gases cannot escape, so they froth up the molten rock and are trapped in the solidified lava. This type of rock, known as 'pumice', is so light it will float on water.

In many volcanic eruptions, the lava and gases are thrown several miles up into the sky as hot clouds of gas and minute glassy fragments, which fall as ash and may become welded together. Huge lumps of molten lava are also hurled skywards and they fall back to Earth as

volcanic bombs, the more liquid of which land and solidify. All the materials that are hurled into the air to fall back over large areas are known as 'tephra'. The debris from the volcanic explosion of the Greek island, Thera, in about 1500 BC spread as far as 800 km away. The centre of the volcano was blasted and

▲ Haytor, Dartmoor, England is part of the upper surface of a granite batholith.

The formation of igneous rocks
1 Upper surfaces of granite batholith
2 Volcanic cone formed by ash and lava
3 Tephra
4 Basalt
5 Main crater
6 Granite batholith
7 Magma
8 Magma
9 Magma

the sea filled the resulting 130 square km of crater.

Volcanic rocks are the results of the swiftest cooling of lava and are so fine-grained that they are susceptible to rapid erosion. If the gas is able to escape, but the lava cools rapidly, a volcanic glass called obsidian results. However, the majority of lavas tend

▲ A tourmaline granite section at a magnification of × 20, showing its large crystals.

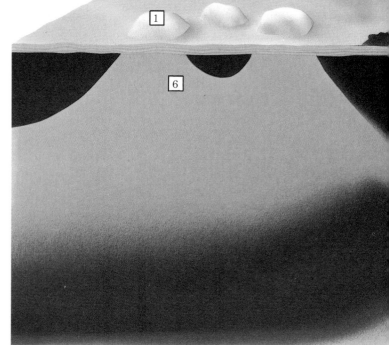

▲ The characteristic hexagonal columns of basalt in the Tertiary lava flow of the island of Staffa.

▲ Molten lava flows down the sides of the volcano, Mauna Ulu, Hawaii.

to pour out over the countryside in thick flows. Because of their volume, they cool more slowly and the different minerals crystallize out to give a texture reminiscent of granulated sugar. As the material in thick lavas cools, it also contracts and, if it is of similar composition throughout, the shrinking and cracking will be even. This characteristically results in the formation of six-sided columns. Basalt is a typical example.

When lava crystallizes before it reaches the surface, the length of time it takes to solidify is greatly increased, allowing longer for the individual crystals to grow. The rock formed is obviously crystalline and it may be possible to recognize that it is made up of different minerals. Some minerals, such as feldspars, come out of solution and crystallize before others, so that a lava can be erupted almost as a crystal mush. On reaching the surface, the liquid part crystallizes immediately. The solid rock will have relatively large crystals of one or more minerals set in a very fine-grained matrix. The tiny crystals of the matrix indicate the speed at which the lava solidified.

When magma rises into the crust, it may melt large volumes of it to form a huge, underground, molten reservoir, called a 'batholith'. The chemical composition differs from that of basalts, which erupt directly from material derived from beneath the crust. The giant batholiths cool extremely slowly, which allows the crystals of the various minerals to grow comparatively large. Because these rocks are formed deep in the Earth, they are known as 'plutonic rocks' after Pluto, the Roman god of the underworld. The most common 'plutonic rock' is granite. It is obviously crystalline, formed from a mosaic of interlocking crystals that have grown into each other. The crystals themselves are very irregular as their individual growth is interfered with by the growth of the adjacent minerals. In some granites there are large crystals of pink orthoclase feldspar, about 2.5 cm in length. These grew before the main cooling and solidification took place in a similar manner, but on a larger scale, as the crystals which occur in some of the lavas, such as basalt.

When the overlying rocks are eventually eroded away, the granites, formed deep in the crust, are exposed to the surface, where, because they are very hard and resistant, they usually stand out as high ground. These huge granite bodies are irregular in shape and one well-known example occurs in south-west England, where its upper surface is exposed in a few patches marked by Dartmoor, Bodmin Moor, St. Austell and Land's End. These separate granite regions are parts of a single batholith still buried deep in the Earth's crust.

Virtually all the sedimentary rocks are derived ultimately from such igneous rocks as granite. The very first rocks were basalts of some sort and the wide variety of rocks on the Earth are the product of the many different ways the basic components of these original rocks were redistributed.

▲ A basalt section at a magnification of × 13. The large crystals formed below the Earth's surface; the surrounding fine-grained crystals formed at the surface.

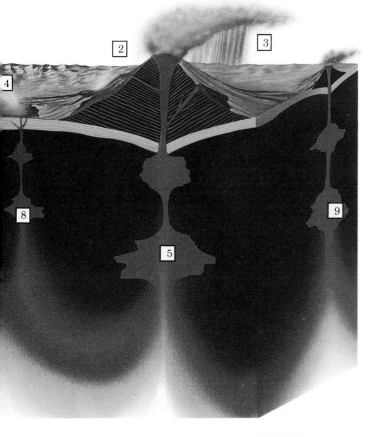

▲ A fountain of red-hot lava has built a cinder cone around the volcano's vent, Kilauea, Hawaii.

▲ Tephra at the edge of the volcanic crater on the Greek island Santorini (Thera).

DESTRUCTION BY PLANTS

Most of the land is covered by plants which protect the rocks from the weather. But plants themselves can physically break up rock. The rock particles mixed with organic matter form soil.

The entire surface of the land is composed of rocks, although sometimes the sediments are not consolidated. Whether they are granite mountains or loose gravels they can be considered as rock in one of its many guises. However, apart from such places as deserts, the Arctic, cliffs and quarries, the main thing to be seen is not rocks but plants. Even when plants have been burned off and the ground ploughed up prior to the sowing of a new crop, there is still no rock visible. Instead, there is a fine brown substance – soil.

The upper layer of earth consists of disintegrated rock fragments mixed together with organic remains. The blanket of plant life is one of the major agents of the destruction of rocks. The classic agents of erosion, the wind and the rain, rarely act directly on rocks but almost always work through the agency of plant activity.

The most obvious method of plant activity in breaking up rocks is dramatically demonstrated by the fate of some of the great temples of earlier civilizations in Central and South America and in Kampuchea, where the jungles have returned and trees and their roots have forced their way through the masonry, disintegrating edifices built for eternity. On a smaller scale, a deserted road or a neglected garden path will be rapidly broken up by plants, even tiny ones forcing their way through the asphalt. The presence of plant roots seeking anchorage and extracting minerals would by itself break up an original rock into smaller units. The process of destruction begins at the very moment a rootlet obtains its first purchase.

It is worth looking for an old road or path and making a record of what has happened to it. It should be possible to find out, from local records or from neighbours, how long it has been left to the mercies of the weather. Former roadways will probably be physically broken up, plants will be growing through the old road surface and, after three or four years, fully grown trees may be established. By making a survey of derelict inner city plots, it is possible to gain an insight into the rate at which plant life can re-establish itself in the urban landscape. For example, during World War II, some city bomb sites were left untended for periods as long as five years, and, by the end of the war, they had a flourishing flora of trees and wild flowers that had become firmly established on the rubble of buildings which had collapsed.

▲ The jungle is gradually enveloping the gateway of Ta Som Temple, Angkor Thom, demonstrating the power which plants have to physically break up rock.

▲ Tree roots have gripped and crushed the masonry of Preah Khan Temple, Angkor, Kampuchea.

▲ Landscape in South Africa showing dramatic erosion. Without plant cover soil can be removed at amazing speed by wind and water.

The role of plants is, however, more fundamental than this mere physical breaking up of rock. When plants die down and when their leaves and petals fall to the ground, they form a plant litter which is gradually decomposed by bacteria and fungi. The decomposed plant matter forms a black, rich organic material known as 'humus'. The process of breakdown produces organic acids which attack the soil minerals chemically. They are also attacked by weak carbonic acids formed by carbon dioxide from the atmosphere dissolved in rain. The percolating waters remove all the soluble materials and very fine particles, such as clay minerals, which are carried away in suspension. Humus, clay, iron and aluminium compounds are thus washed out of the upper layer of the soil, a process called 'leaching'. Beneath the plant litter there is a grey or white leached layer, known technically as the 'A horizon'.

The chemicals removed from this layer are carried downwards and are redeposited in what is known as the 'B horizon' or accumulation horizon, because it is here that the nutrients carried down from the upper layers accumulate. The precipitation of iron compounds, and humus in particular, gives a characteristic appearance to this layer of the soil. The upper part is brown or black because of the humus, while beneath there is a yellow or red layer rich in iron oxides and clay. If this zone is left undisturbed, it may form an impervious iron-enriched layer or pan. Beneath this is a sort of transitional zone of partly broken up rock, which is only slightly altered both physically and chemically from the bedrock. This is the subsoil or 'C horizon'. Finally, the unaltered rock is reached. This may be any type of rock, such as a limestone, a clay, sandstones or gravels.

The pioneering soil studies were done in Russia, so all the main types have Russian names; the most common one, just described, is 'podzol', which is found in all the temperate latitudes with a moderate rainfall. This includes most of Canada, excluding the Arctic, and the eastern half of the United States, western Europe as far as the Urals, Siberia and China, south-west Australia and the eastern part of New Zealand.

An obvious disadvantage of podzols is that all the nutrients are leached out of the upper layers. The reason fields are ploughed and gardens dug is because this brings up the nutrients in the accumulation layer to make them available for the plants. However, ploughing and digging regularly remove the protective plant cover and expose fine loose particles of rock and organic matter directly to the agents of weathering. Consequently, when the soil has dried out, wind will readily blow it away and any sudden downpour will wash it away. To lessen the effect of removing the plant cover, farmers plough along the contours rather than up and down a hill, which would result in all the soil being washed to the bottom of the hill.

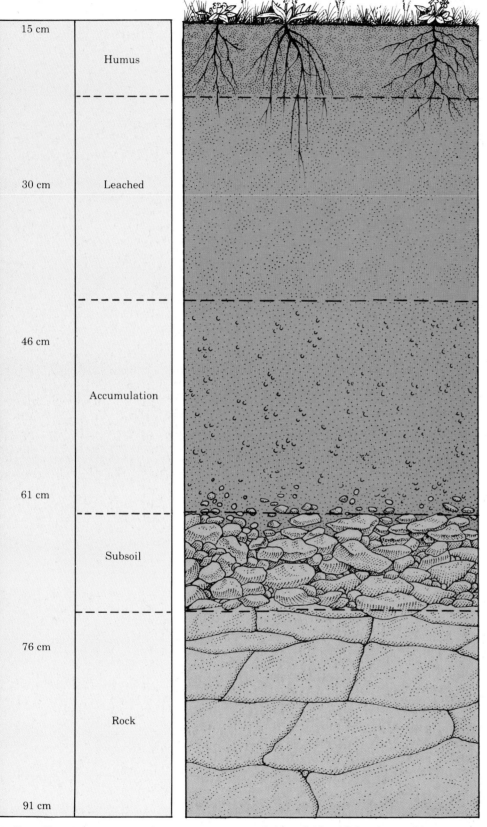

15 cm	Humus
30 cm	Leached
46 cm	Accumulation
61 cm	
76 cm	Subsoil
91 cm	Rock

Soil profile
One of the best ways of discovering how much the underlying rocks have been transformed by the deep weathering of soil formation is to examine any vertical section in the ground – a soil profile. To do this dig a hole ½ – 1m deep, making sure the sides are kept clear and sharp. It should then be possible to determine how far the soil has been modified by the activity of man. In most cultivated areas, the leached and accumulated levels have been churned up but, if the ground has not been cultivated, there is a good chance of finding a fully developed original soil profile.

CHEMICAL DESTRUCTION

**The chemical breakdown of rocks is part of
nature's recycling process. The erosion it causes can be the
first step in the formation of sedimentary rocks.**

Chemical weathering is one factor which contributes to the breaking down of rocks during the geological cycle. This type of weathering changes the composition of rocks and allows them to be more easily transported. For chemical processes to take place it is usually necessary for water to be present. Water reacts with the minerals in the rocks and may even dissolve them; it can also contain dissolved gases which combine with it to form acids which attack the rock.

Many of the minerals that have crystallized from molten rock to produce large crystals as in granites, or the minute ones in such fine-grained rocks as basalts, are less stable at surface temperatures and pressures and therefore are susceptible to chemical changes. Feldspars are the major constituent of all igneous rocks. This family of minerals is composed of aluminium silicate, fundamentally similar in chemical composition to silica, but containing a small metallic ion of potassium, sodium, calcium or barium (the last is very rare). During crystallization calcium feldspars form first, then the sodium and then the potassium; the latter two are the most stable. When feldspars come in contact with water, a chemical process known as 'hydrolysis' (which means breakdown by water) occurs. The feldspar's metallic components, such as calcium or sodium, combine with some atoms in the water molecule, often forming a highly soluble compound. The sodium ends up as sodium chloride or common salt. Calcium eventually becomes calcium carbonate, which makes up limestones. The replacement of the metallic atoms by hydrogen from the water produces complex hydrous aluminosilicates or clay minerals which are arranged in platy sheets. These break down into minute flakes that are carried away in suspension as clay particles. The particles will eventually be redeposited to form clay, mudstones or shales which account for 45 per cent of all sedimentary rocks.

When limestone is exposed to the atmosphere, it is vulnerable to a different type of chemical attack. As rain falls through the atmosphere, the carbon dioxide in the air combines with some of the water to form a weak carbonic acid. (If you blow air into water a weak acid will form, because air that is breathed out contains carbon dioxide; this acid should turn litmus paper red.) When carbonic acid comes in contact with calcium carbonate (limestone), some bicarbonate is formed; as this is soluble in water, part of the limestone will dissolve away.

▲The limestone pavement at Malham, Yorkshire, England shows that there is greater erosion in a humid area.

▲ The limestone columns of this Greek temple show that the rate of chemical erosion has been relatively slow in the dry climate.

▶ These limestone pinnacles in Sarawak represent an extreme example of erosion in tropical rain forest conditions.

A similar process is used by bird and reptile embryos: as they respire they produce carbon dioxide, which forms carbonic acid which dissolves a small amount of the calcium carbonate in the inner surface of the eggshell; the embryo is then able to make use of the calcium in the formation of its internal bony skeleton.

The most obvious example of the acidic erosion of limestone by solution is seen in the formation of cave systems (see pp 54–5). Among the more dramatic types of chemically weathered limestone are severely eroded marble statues and gravestones (formed wholly or partly of limestone); from their dates we can measure the rate at which they have been eroded. This is greater in industrial areas because smoke containing sulphur dioxide changes in rainwater to sulphuric acid, which is a very strong acid capable of dissolving carbonate very rapidly. Chemical weathering also attacks the mortar (a mixture of lime, sand and water) used in cementing bricks. The mortar gradually disintegrates, hence the need for repointing houses at regular intervals.

The significant feature of chemical weathering is that the minerals which go into solution are then in a condition to be easily transported elsewhere. Most natural water has calcium carbonate dissolved in it. If water in a glass is left to evaporate (a process that will be accelerated if there are flowers in the glass, because they draw up water which evaporates from the leaves), a thick white crusty rim of calcium carbonate will be deposited on the inside of the glass. This will be difficult to remove by washing or scraping, but a weak acid such as acetic or vinegar will dissolve it again. The water that runs down the walls of caves similarly deposits calcium carbonate as it evaporates, thus forming the so-called 'flowstones', because they give the appearance of having flowed over the surface of the walls.

Perhaps the most familiar examples of chemical erosion involve the process known as 'oxidation', where oxygen combines with a mineral or element and in so doing changes its state so that it is more easily recycled. For example, when coal (carbon) is burnt, it is oxidized to carbon dioxide, which is returned to the atmosphere. It can then be used by plants to synthesize sugars and other carbohydrates, which can, given sufficient time, be converted back to coal. Indeed hydrocarbons, such as oil, and plastics, such as polythene, are all oxidized to carbon dioxide and water and hence are recycled. Unfortunately the rate at which this takes place is not rapid enough to prevent them adversely affecting the environment in the short term, well illustrated by the pollution caused by oil spills.

One example of oxidation which can be easily observed is the recycling of iron by rusting. Watch a nail left outside for several days: in the presence of water the iron will combine with oxygen to form limonite, a hydrated (containing water) iron oxide ($FeO.OH$) which readily goes into solution. When this dries out and the water is evaporated, it gives rise to ferric (iron) oxide (Fe_2O_3), which has a bright red colour called 'haematite' (from the Greek word for blood). When a nail rusts, or a car, or a bolt, the adjacent area is stained red. This means that the iron has gone into solution, been transported and been redeposited so that it is recycled.

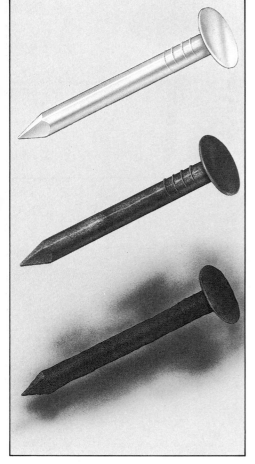

▲ As a nail rusts, the iron is oxidized and the resulting iron oxide is deposited around it as red ferric oxide.

▶ Steel girders rusting in the open air show the same chemical destruction at work.

PHYSICAL DESTRUCTION

Rock is initially broken up physically by friction or by the effects of heat and cold. Wind and water then complete the process of physical destruction.

The most obvious way in which an object can be destroyed is for it to be broken into pieces. The material will not be altered chemically, but simply reduced to smaller and smaller fragments. This process can be the result of sudden impact or of repeated collisions of objects which chip the edges off one another to become smooth and reduced in size.

If one starts with, for example, an igneous rock like granite, some of the minerals such as the feldspars and micas undergo chemical changes and are removed either in solution, or in suspension (see pp 16–17); this leaves the glassy constituent quartz, which is the crystalline form of silicon dioxide or silica. Silica is very hard and chemically inert; this is because of its structure. The silicon atom is attached to four oxygen atoms in the form of a solid triangle or 'tetrahedron'. These are packed so closely together in a crystal lattice that the oxygens between adjacent silicon atoms are shared, resulting in only two oxygen atoms for each one of silicon (SiO_2). The silica is loosened and isolated from the rock by the chemical destruction of the intervening minerals; this exposes the quartz to physical destruction.

Where the granites have rotted chemically, for example in the china clay mines in south-west England, the feldspar is changed to the clay mineral kaolinite, or china clay. Simply by turning high power jets of water on the rock the kaolinite can be removed, leaving behind vast piles of quartz crystals, which stand as large, brilliant white tips dominating the landscape. The angular irregular quartz crystals can be carried away by water, and as this happens, the crystals continually strike against each other, so that their sharp edges are smoothed away. This is described as rounding and is applied only to the surface of the grain. Its volume is described in terms of its sphericity, in other words, its deviation from a sphere of equivalent volume (see pp 26–7).

The break-up of a rock consisting of different types of mineral can also be accomplished by their varying reactions to heat and cold. If one mineral expands more than another when heated, the rock will develop lines of weakness along which it will subsequently fracture. One of the major forces in breaking up rocks is the freezing of water in the rock pores or fractures. The ice crystals take up more room than the same amount of water in its liquid state. This expansion pushes particles of rock outwards, so that when the ice melts, the fragment falls straight on to the ground. On a slope the particles would

▲ Earth pillars in the Hoodoo River Valley, Yoho National Park, Canada.

▲ The formation of earth pillars. Vertically falling rain washes away loose sediment, while the large stones act as umbrellas to protect the ground immediately beneath them.

gradually move downwards.

An analogous situation occurs in hot desert regions, where the growth of salt crystals has the same effect of forcing apart pieces of rock. Once a rock has suffered the effects of being broken up, or if it is composed of sediments that are unconsolidated, wind and water can complete the process of disintegration.

Falling rain is capable of dislodging particles and as the water flows downhill, they will be carried away. Where rain falls vertically in a narrow or sheltered valley, or by the side of a sunken road where soil is exposed, and the sediment contains large pebbles or boulders, these remain while the sediment around them is washed away. The boulders provide a protective covering for the soft sediments immediately beneath them. This situation results in earth pillars, which are very striking landforms and give a clear indication of the amount of erosion caused by the simple action of falling rain.

When rainwater collects into streams and then rivers, a further kind of physical destruction comes into play. During the turbulent flow of waters (especially during floods, say after the spring melt of snow), blocks of dislodged rock are tumbled downstream and, together with smaller fragments and sand, break up and reduce the size of the boulders in the river bed. Where boulders become caught in a depression on the bed of a river or stream, they will be swirled round and wear out a deep hollow or pot-hole in the bed of the stream, rather like the action of a pestle and mortar. The effect of the power of rivers in breaking rock can be seen where the land has been uplifted and the rivers have cut down through the rock to keep pace. A classic example is the Grand Canyon in Colorado, in the United States.

The other major agent of physical destruction is the wind and, as with water, it can only work on rock after the initial breakdown has been accomplished or if the sediment is not consolidated. In the desert, differential heating and cooling and the growth of salts or ice crystals lead to the weakening and disintegration of the original rock. The wind then removes the finer particles, leaving behind a rocky or stony desert. Clay minerals are light enough to be carried vast distances, but grains of sand, which are heavier, are bounced along; this type of movement is known as 'saltation'. As they bounce, their surfaces are smoothed and the grains themselves become spherical: each individual grain is sandblasted and itself sandblasts others. Since wind is not able to lift particles of sand much above 2 metres, the effect of sandblasting, which facets and erodes exposed rocks, is capable only of destroying rocks up to a height of 2 metres.

Although it is possible to consider the physical and chemical aspects of erosion separately, in reality both processes accompany one another. They combine to break down the rocks so that they are then ready for the next stage in the cycle: transportation.

▲ Cathedral Gorge, Nevada, USA is a good example of badlands erosion, caused by occasional torrential rain in an arid area.

▲ Three-sided stones known as dreikanters, which have been faceted by wind-blown sand.

▲ These are frost shattered fragments from a dark boulder on a Norwegian beach.

PEBBLES

The presence of pebbles on beaches and in rivers provides evidence of distant environments. When pebbles become cemented together they form conglomerate rock.

One of the first tangible signs of the erosion of the land, the transport of material and its deposition is the presence of pebbles. Pebbles are fragments of rock that have been rounded and smoothed; they form the basis of the sand and gravel, or aggregate, industry; gravels are extracted on an ever-increasing scale for the requirements of the building industry. Concrete is an artificial conglomerate made from a mixture of sand, gravel and cement. The mixing is in a confined space so that the individual components will not separate out from one another.

In the initial process of being transported by streams and rivers, as opposed to being swirled around in a cement-mixer, the products of erosion sort themselves out to form two main types of rock: 'breccias' and 'conglomerates'. A breccia contains rock fragments which have only travelled a short distance. It is characterized by the sharp, angular fragments produced by the initial fracturing of rocks. For example, when frost-shattered rocks which have simply tumbled down a mountain-side to make a massive scree are consolidated, the rock is termed a breccia. Consolidation can occur through deposits of minerals such as calcium carbonate or iron oxide, or mud, cementing the fragments together. Sometimes fissures can occur in a rock due to earth tremors and are then cemented by mineral deposits. Although the rock has remained *in situ*, it is a breccia because it reveals evidence of once having broken up. When broken angular fragments are swept away suddenly by flash floods or mud slides, they too form breccias.

Breccias are always a sign of movement over relatively short distances. When the distance over which the rock fragments travel increases, the sharp edges are rounded off and the resulting rock is termed a conglomerate. There is no certain distinction, but with any appreciable sign of smoothing, it is best to call the rock a conglomerate. Conglomerate rocks are consolidated in the same way as breccias.

The rounding of boulders and pebbles as a direct result of repeated collision takes place very quickly. In transit the materials are sorted out, with the smallest particles – sand grains and silt – being carried further than the pebbles.

The best place to study pebbles is on a shingle beach. The first task is to find as many different types of pebble as one can. Usually the pebbles will not be derived from rocks in the immediate vicinity. In most cases they will come from elsewhere and will have been moved along the coasts by the action of tides. The waves usually strike the coast at an oblique angle, but recede at right angles to it so that the sand and pebbles washed in by the waves follow a zigzag route along the

▲ Budleigh Salterton pebble bed, Devon, England. In the foreground are pebbles

derived from the cliff by wave action to form a modern pebble beach.

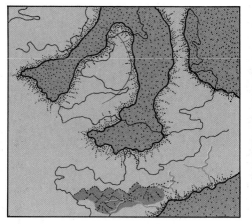

▲ Map of Triassic geography showing derivation of the Budleigh Salterton pebble bed from the ancient mountains of Britanny.

▲ Large scale cross-bedding in the Budleigh Salterton pebble bed.

beach; this is known as longshore drift. Pebbles in a river gravel will also provide evidence of their source and give an idea of the size of the catchment area. Conglomerates occur in two major settings – in river systems and along coasts.

Two kinds of conglomerates can be distinguished from the way in which the pebbles lie: either the pebbles mainly rest on top of each other in direct contact, or they are separated from one another by sand or smaller particles.

On a shingle beach the pebbles are in direct contact with one another and as the waves break and retreat, moving the pebbles about, the sand grains are winnowed out from between the pebbles. If a beach is examined in section, there will be patches where layers of sand lie within the layers of pebbles, but in the main the different grades of sediment will have been fairly well sorted. One of the features of beach deposits is that the angle of rest of the shingle is always low – about 10°, which is not far from the horizontal when viewed from eye-level. A sorted deposit of pebbles of this nature can generally be identified as a fossil beach or offshore shoal.

In a river environment, the nature of the conglomerates formed by river action contrasts with those of beaches formed by wave action. In a meandering river system traversing a flood plain, the main deposits are channel sands, but pebbles are dragged or rolled along the river bed. As the meander moves laterally and downstream, channel sands are deposited on top of the pebbles and a thin conglomerate is left behind. These conglomerates, known as bottom lag deposits, are not the immediate consequence of erosion from a nearby source. They consist of sands and individual pebbles, usually of a varied rock type, from several sources, collected from the different tributaries of the main river system. Detailed studies of the rocks from which the pebbles derive allow accurate accounts to be given of the sources of sediment from the distant hinterland.

The other type of conglomerate from river systems is where floods have swept boulders and cobbles in raging torrents and dumped huge masses of unsorted sediments. One of the classic examples of such a deposit occurs at Budleigh Salterton in Devon, England, which is famous for its pebble bed of Triassic age (about 225 million years ago). Most of the pebbles are composed of a quartzite, which is a sandstone cemented by silica. They are embedded in a matrix of coarse sandstone. There has clearly been minimal sorting, which suggests that the sediments were deposited rapidly, perhaps dumped at the foot of mountains. There are no rocks known in Britain from which these pebbles could have derived; however, the right type of rocks is known in Brittany, France. It was the erosion of mountains that produced these pebble beds (see diagram). As the present-day cliff has been eroded by wave action, the pebbles have been washed clean and form the shingle of the beach. In this one environment it is possible to observe the marked contrast between a sorted beach gravel and the unsorted gravel deposited by a river in spate.

▲ Dry river-bed containing boulders at Pikermi, northern Greece.

▲ A section through a Pleistocene beach deposit, Westleton Beds, England.

▲ Detail of breccia used as masonry at Mycenae, Greece.

▲ Blocks of various conglomerate rocks were used to build the ancient city of Mycenae.

HOW ROCKS TRAVEL

**When rocks erode, the particles are carried away by
wind or water. The way in which they are deposited determines
the characteristics of the new rock they will eventually form.**

Before rocks can be formed from the products of erosion, the sediments must first be carried away from their place of origin. During the journey the sediment is sorted out and deposited, starting the rock formation cycle again. A journey can be over vast distances, for example rock fragments from the Andes are carried across South America to be deposited on the Atlantic coast.

The major agent responsible for transporting sediment is water, generally in river systems. Material derived from a granite, for example, is carried away by three methods: by suspension, dissolved in solution, or by saltation (see pp 18–19). The different materials in the sediment will be redistributed so that the salts in solution will be carried furthest; mud in suspension will be carried almost the same distance, but the grains of silica quartz will not be transported as far. Any major river in the world provides evidence of clay minerals carried in suspension: every river in its lower reaches is muddy and where it meets the sea the salt causes the clay minerals to stick together, to

'flocculate', so that they sink to the bottom to become mud. The larger particles carried by rivers include pebbles which are rolled along the bottom of the main channels and sand grains which are bounced along in a fast-flowing current. Most of the grains will be carried by the water, but as the river reaches a flat alluvial, or flood plain, the rate of movement is drastically reduced and the particles are deposited.

In a mature river system, the river tends to wander or meander across the landscape; as the water flows along, the current is faster on the outside curve, causing erosion, whereas it is weaker on the inner part of the curve, causing sediment to be deposited. As the river moves across the plain, it leaves behind a layer of pebbles and a series of sands, known as 'pointbar' or channel sands. Because of the meander, these deposits are built up at a marked angle to the horizontal, so that the rock that is eventually formed is sloping; this is known as 'cross-bedding' and the slope indicates the direction of flow. On top of the cross-bedded sands are sets of much

smaller scale cross-bedding, known as 'cross-lamination'. In section this resembles a series of ripples, while cross-bedding appears more like a series of small dunes. The reason for this difference in scale is that as the water becomes shallower, the flow weakens and so the sand deposits are moved more gently.

When sand particles are moved by either wind or water, they are piled up until the heap becomes unstable and the lee face (the steeper side) avalanches, stopping when it reaches a stable angle of rest. The small heap of sand caused by the avalanche is built up by more deposits until it avalanches and the process is repeated.

Once this pattern has been established, it is maintained by the current itself. After the current has passed over the crest of the ripple,

▲ Sand-dunes in the Great Oriental Erg (sand sea) near El Oued, Algeria.

▲ An aerial view of a river mouth showing sediment being deposited.

▲ A salt lake where salt is coming out of solution, another form of deposition.

there is a reduction in pressure due to the subsequent dip. A small reverse current or eddy is drawn in, contrary to the main flow; this pushes against the lee side, helping it to maintain its position.

As a result of recent studies, it has become accepted that the nature of the structure of sand bodies is a direct consequence of two separate and independent factors: the size of the particles and the rate of flow of the current. As the particle size can be observed, it is possible to calculate the rate of flow.

When a rock is exposed, as on the coast, one can see the repeated sequences of alluvial deposits from which it was formed: first a layer of conglomerate, followed by cross-bedding, then cross-lamination covered by a layer of silt or mud. The sequence is then repeated and can continue for hundreds of metres. Each sequence of layers represents a river deposit, and shows, moreover, that the particles deposited gradually decrease in size. This can be explained by using as an example cross-bedding giving way to cross-lamination. The power of the

▲ Stalactites and stalagmites show that limestone is transported and redeposited.

▲ Stalactites and stalagmites can eventually meet to form solid columns. The Veiled Statue, Carlsbad Caverns, New Mexico, USA.

stream flow weakens as it moves up the slope of the inner curve of a meander; this is due to the relative increase in surface area in contact with the water, which causes increased friction or drag. This slows down the water flow, which causes smaller ripples. Thus the size of the particles in a rock layer can tell us how strong the river current was.

The same approach can be used in studying the deposition of sands in deserts, which also forms ripples and dunes, although there is a further set of structures known as 'dra', which are sand dunes on a truly gigantic scale. The feature of sands deposited by wind is that their angle of rest is appreciably steeper than those in water. The sand grains too are better graded. If one sees cross-bedding with smooth and spherical grains, it is usually the result of wind deposits.

The flood plains of rivers can be covered by fine silts or muds. These are deposited when rivers burst their banks during floods and turbid muddy waters spread over the land surface. When the water retreats it leaves behind a layer of mud. This is one of the major ways in which clays and muds are deposited in a terrestrial environment. Where clay minerals reach salt water, they are deposited by virtue of the flocculation caused by the salts.

Minerals transported in solution can be deposited in a variety of ways.

The most obvious can be observed on land, in particular in desert regions where the water-table is close to the surface. Groundwater containing minerals in solution is drawn up to the surface where it evaporates, leaving behind minerals like calcium sulphate or gypsum. These minerals can occur in attractive crystal shapes, forming what are known as 'desert roses'. Many salts are deposited by evaporation. For example, in many coastal areas, fresh groundwater evaporates and is replaced by salt water drawn into the soil pores; when this evaporates, salt is deposited, as in the Persian Gulf sabkhas.

The other means of mineral deposition from solution is found in shallow warm seas under the influence of tides. Here calcium carbonate is deposited on small particles of sand or broken shell. As they are washed back and forth, they produce small spherical grains or oolites (from the Greek *oion*, an egg), hence the term oolitic limestone which is used as a building material. It has a rough grainy texture rather like cod's roe.

A major way in which minerals are deposited is through the agency of a living organism. The most familiar organic deposit is chalk, which actually consists of the skeletons of microscopic one-celled marine algae. Numerous other limestones are made up of animal skeletons, either as corals or shells (see pp 58–9).

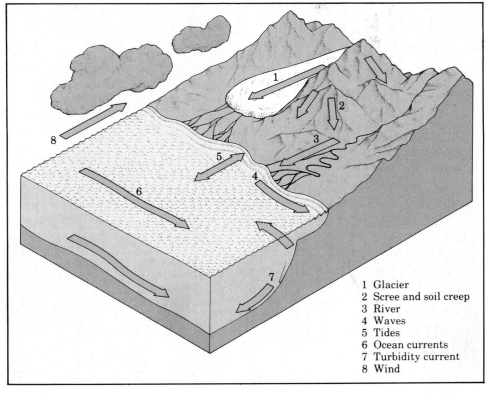

1 Glacier
2 Scree and soil creep
3 River
4 Waves
5 Tides
6 Ocean currents
7 Turbidity current
8 Wind

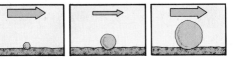

◀ The diagram shows that a very strong current is needed to lift both very tiny and large particles. A weaker current will move ordinary grain-sized material.

HOW TO RECOGNIZE PAST ENVIRONMENTS FROM ROCKS

**If we look carefully at the rocks around us, we can
tell whether they were once deposits in the ocean, sediment from
river systems, or dunes in the desert.**

Sedimentary rocks are a record of past conditions. Examining the details of the sediments, together with the fossil remains of animals and plants, can indicate the environment in which they were formed.

The sediments that are the easiest to interpret are those deposited in the sea near the coasts on continental shelves. These sediments, whether sandy or muddy, contain direct evidence of living things, usually in the form of shells, or burrows and traces of marine creatures. Where little or no sediment enters the area from the land, calcium carbonate can readily come out of solution and more calcareous deposits are formed. These may consist entirely of microscopic shells of animals and plants or may be the result of the breaking up of shells or other skeletal remains that have been swept up into shell banks.

Coral reefs too can be easily interpreted since corals only grow at a certain temperature and depth in clear water. Reefs are therefore evidence of these conditions.

One can see abundant evidence of deposition in shallow seas from fossiliferous or oolitic limestones. Often oolitic limestones display cross-bedding, and there is also evidence of eroded channels. But these cross-beds were not formed by rivers because cross-bedding alone is present and not the sequence of layers built up by rivers (see p 23). However, they must have been deposited in an environment where there were strong currents and

channels. In fact, it is not unusual to find fine sands showing what are called herring-bone cross-beds, where the cross-beds slope in two different directions so that they resemble the bones of a herring. This type of cross-bedding shows that there were strong currents depositing sediment first in one direction and then in the opposite direction. Such conditions occur in areas which are subject to the action of tides.

In many regions there are thin layers of sand interspersed with thin

layers of silts or clay, with little sign of living things preserved apart from occasional shallow burrows. This repeated sequence of sand and mud is almost always the trademark of intertidal flats, which contain few organisms apart from those that burrow in the sediment. One of the important clues in these deposits is the presence in the sand layers of ripple marks which are symmetrical. This is evidence that the area was not subjected to the action of currents flowing in a single direction, in which

▲ Fossil Triassic desert dunes in Devon, England.

▲ Cross-bedded sands of desert dunes in Zion National Park, Utah, USA.

▲ Fossil ripples in Utah, USA formed along the margin of a lake.

▲ Cross-bedded sandstone from a river deposit in southern Tunisia.

case the slope of the ripples would show the single direction of flow.

Occasionally we come across large-scale cross-beds that were formed by dunes up to 10m high which seem to be of the same dimensions as some of the larger desert dunes, both of the present and the past. Yet when they are examined closely, they are seen to be composed of alternating layers of broken seashells and coarse shelly sands. There is no possibility of such structures being constructed on land by the action of winds. In fact in shallow seas, such as the North Sea at the present time, underwater dunes of sands and shells are built up by marine currents. Exactly as the smaller-scale ripples and dunes form in the river bed, so do dunes on the sea-floor. Loose particles and a moving current, be it a marine current, a wind or a river, produce the same basic structures, which then travel in the main direction of the current. The discovery that marine dunes slowly move over the sea-floor, just as desert dunes migrate, is of considerable importance in areas where offshore oil drilling is envisaged. One important result of the experimental study of current flow and the types of ripples and dunes that form is that height seems to be a factor of depth, so that the height of a dune is approximately one sixth of the depth of the water in which it was formed.

Cross-bedding giving way to cross-lamination is the sign of a deposit in the flood plain of a meandering river (see pp 22–3). Where such river channel deposits have been succeeded by fine silts or muds, a period of flooding is indicated. If this layer of mud shows cracks, it must have been exposed to the air; the drying out of the flood waters caused cracking in the uppermost layers of mud. When a rock forms, these cracks are filled in by another layer of sediment.

In some regions limestones and mud alternate, with little evidence of life apart from thin-shelled freshwater snails. This is proof of the former existence of freshwater lakes which have been filled in by the sediment from the river which fed them. Because they cannot exist in any other environment, the presence of some types of pond snail establishes that the water was still, while other snails indicate flowing water. There are some types of limestone which have a rubbly texture and are devoid of fossils. These were not formed under water, but within ancient soils by the evaporation of water in the soil pores and the deposition of calcium carbonate. These limestones are called caliche or calcretes. As nothing could exist in these conditions, fossils, apart from roots, are never found in calcretes.

Sandstones, in which the individual grains are spherical, rounded and very well graded, indicate transport by wind. Such sands in giant cross-beds with angles of rest up to 30° show the rocks to be fossilized desert dunes. Associated with such deposits are irregularly eroded surfaces, sometimes with undercut channels, filled with angular fragments. These are formed by erosion due to the violence of flash floods which, when the flow ceased, simply dumped unsorted angular fragments in the channels which the floods had just cut.

Although each one of these major types of environment can be analysed in greater detail, it should be possible to recognize them by examining the different types of sedimentary rocks.

▲ A fossil marine dune at Bawdsey, England.

1 Surface ripple
2 Cross-bedding
3 Current
4 Eddy
5 Avalanche edge

▲ The appearance of ripples on the surface and their cross-section appearance, or cross-bedding, which can be seen fossilized in rocks.

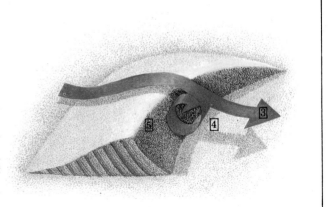

▲ The formation of a small ripple or sand dune. The air or water current pushes up the particles until they reach a high point and collapse. As the current flows over the avalanche edge a small eddy forms which helps to maintain the structure.

HOW SEDIMENTS BECOME ROCKS

After deposition, loose pebbles and sand grains have to be consolidated to form rocks. Generally, they are glued together by mineral deposits or compacted by the weight of overlying sediments.

When sand is dumped in an alluvial plain or carried to the mouth of a river and when silts and muds accumulate on the sea bed, they are mixed with water and are unconsolidated. One of the first changes that must take place in the transformation of sediment into rock is the removal of water. There are two basic ways of accomplishing this: either the water is squeezed out or it is replaced by some other material.

To make rock from mud, which comprises clay minerals and water, the water is usually squeezed out, although sometimes it simply evaporates. The clay minerals are flat or platy and have the property of attracting molecules of water to themselves. In a mud, the clay minerals do not have any particular orientation so that if the sediment simply dries out, the resultant rock will be a mudstone which, when it fractures, will break into irregular blocks because of its homogeneous nature. Such a rock will be impervious to water.

Most muds, however, are transformed into rock as a result of compaction by the weight of overlying sediments. The different degrees of compaction can be best appreciated by looking at clays and shales of different ages, where the hardest and oldest rock would be the deepest buried. However, in some regions, such as the Baltic, very ancient clays remain relatively unchanged because they were not buried beneath thick sediment.

The first stage in forming rock out of mud is when most of the water is squeezed out by compaction, thus producing a clay. This is a soft plastic type of rock which can readily reabsorb water and return to the consistency of mud. This is especially the case with many Mesozoic and Tertiary clays that absorb vast amounts of water and then move as mudflows. This is a common occurrence after heavy rains and can make fossil collecting an extremely hazardous undertaking.

With further compaction, still more water is forced out of the clay and the platy minerals become aligned parallel to the beddings. Such rocks are shales and these fracture along the bedding planes, unlike clay where the mineral particles are not aligned. Where such shales are subjected to intense pressure they can be further changed to slate, although in this instance the clay minerals are aligned perpendicular to the force causing the pressure and this may not bear any relationship to the original bedding. The slate will then split along the plane in which the minerals have aligned, which is called the 'cleavage plane'. Slate is very hard and impervious to water, which makes it a good roofing material.

When sand is first deposited, there is water between the grains, which has to be replaced for the sand to become rock. Sand with water surrounding the grains can be transformed into quicksand, which is a liquid, simply by agitating it; any struggling object caught in it can then be engulfed.

Sandstones and conglomerates

A series of photographs of sand grains taken under a scanning electron microscope to show very fine details of their surfaces.

▲ A grain with sharp, jagged edges which has been transported a short distance.

▲ The edges have been rounded by further transport.

▼ **How sand and mud form rocks**

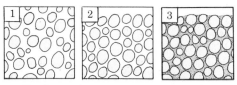

▲ 1 Sand particles are deposited loosely. 2 With pressure they pack together more closely. 3 Rock forms when they are cemented together.

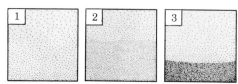

▲ 1 Clay minerals and water form mud. 2 The particles settle. 3 The water is squeezed out by compaction from the overlying sediments and rocks form.

▶ An overgrown grain.

are too hard to be compacted by weight. Where sediment has been rapidly deposited, the grains will be loosely packed. However, with pressure and in particular with a shaking movement as in earthquakes, the grains can be more effectively packed. This can be compared to pouring sugar or coffee into a glass so that it fills to the brim; when the glass is gently tapped, the individual grains rearrange themselves more efficiently.

However closely the grains of sediment pack together, if rock is to be formed, the loose sediment has to be cemented by deposits in the spaces between the grains. This can happen very easily, as seen, for example, on beaches when 'beach rock' is formed. This is where broken scree or patches of shingle become cemented by calcium carbonate or iron oxide deposited between the fragments. When water reaches the broken rocks or shingle on a beach it soaks into them and then evaporates; in so doing, calcium carbonate or iron oxide is deposited onto the pebbles and rock fragments and cements them together to form a solid breccia or conglomerate. This process can take place in a matter of days, but only occurs in irregular patches. This is an example of a rock being formed almost instantaneously.

The mineral deposited in the spaces between sand grains is termed the 'rock cement' and frequently is of a mineral different from the sand itself. The most common cementing minerals are calcium carbonate and iron oxide. Iron oxides in solution percolate downward through certain soils, like podzols, to reach a depth where they come out of solution and an iron pan is formed. In thick layers of sand the iron will form a matrix in irregular patterns or may form cubes containing loose unconsolidated sand inside.

In contrast to iron, calcium carbonate tends to form around a nucleus, like a fossil, so that sands may be cemented in small isolated patches. In the case of sediments made up of muddy limestones or limy shales, calcium carbonate may come out of solution in the form of large nodules or concretions. These are especially important if they form around fossils because they will prevent the fossils from being crushed flat. Fossils found in shales are compressed, often almost out of recognition.

Sometimes the cement is of the same mineral as the sediment; for instance, oolitic limestone is cemented with calcium carbonate. Sand grains cemented by silica produce one of the hardest rocks – a quartzite.

Sand grains begin as irregular quartz crystals which become separated from other minerals and are then subjected to physical modification. When they are bounced along a river-bed they remain angular, but the sharper edges become rounded. On a beach this process is continued; only in a desert will sand grains become perfectly rounded and spherical. Silica is transported in solution through the sand. When it comes out of solution it crystallizes in the same structure as the sand grain. If it is deposited on the grain, it simply becomes another crystal added to it. In this way a quartz crystal is built up. If the grains are closely packed, the growth of the individual grains will interfere with one another exactly as they did in the original granite from which they came (see pp 12–13). When quartzites are examined under a microscope in polarized light, they will appear as a mass of interlocking crystals. Only in ordinary light can one discern the ghostly outline of a grain of sediment.

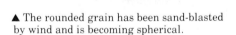
▲ The rounded grain has been sand-blasted by wind and is becoming spherical.

▲ The smooth facets are crystal overgrowths where cementing material has been added onto the original crystal lattice of the grain.

◄ Limestone concretions in the National Fossil Park, Himachal Pradesh, India, which are caused by calcium carbonate being deposited around nuclei in the sediment.

THE UPLIFTING OF THE ROCKS

**Features in our landscape today can tell us whether
it was once part of the ocean floor raised by volcanic forces,
or an ancient river submerged by melting ice-caps.**

There is abundant evidence in the landscape to show that there must be some mechanism capable of raising sediments deposited on the sea-bed to make dry land. This is a reasonable inference in, for example, the case of incised meanders. When water flows down a steep slope, it travels in a fairly straight line, but on an almost flat surface it wanders about; hence, rivers traversing a flood plain close to sea-level will meander. The presence of river meanders is taken as a sign of a mature landscape that has been eroded almost to sea-level. When such meanders are found cut deep into the landscape, it indicates that the land has been raised up. This must have been at a rate slower than that at which the river was able to cut down its bed, otherwise the meanders would not exist.

Around many coasts of the world there are platforms cut out by waves, cliffs and beach deposits with marine shells which occur tens of metres above present-day beaches; these too are taken as evidence of the coastline having been raised. The existence of raised beaches could theoretically be the result of the progressive lowering of sea-level. Evidence that raised beaches are actually the result of uplift can be confirmed by examining the nature of the streams that flow to the sea; where land has been raised, the streams are very straight and cut deep V-shaped valleys or ravines which eat their way back into the land. Where the land has been drowned, the lower

reaches of river systems have been inundated by the sea, and the inlets along the coastline, such as the north coast of the Isle of Wight, England, are simply the outlines of the pre-existing river systems. Many land areas demonstrate both phenomena in close proximity, which means that parts of the land surface are capable of being tipped slightly so that as one part is depressed, the other is elevated.

However, the fact that the surface of the land, including the highest mountains such as Mount Everest, were formed from sediments deposited at the bottom of the sea, or at least close to sea-level, provides incontrovertible evidence of uplift.

There is further evidence of the effects of uplift. Incised meanders such as in the Grand Canyon, in Colorado, where a river system has cut its way down through horizontal layers of rock to a depth of some 1.5 km, demonstrate that the land has been gradually lifted vertically without any appreciable tilting so that the original rock strata still remain horizontal.

When rock strata are examined in many parts of the world, they are not always horizontal, but can be crumpled or folded. Sometimes the folding is very gentle. This implies that there must be some kind of horizontal component in the process of uplift, so that together with being lifted the

▲ A series of raised beaches at Dunure, Ayrshire, Scotland.

▲ Mount Nilgiri, Nepal which has horizontal strata near the summit showing that the entire mountain has been lifted up from the seabed.

rocks are squeezed or pushed from one side. The evidence of this folding on a small scale can be seen in cliff faces or quarries where the strata are crumpled. In some cases they may be single beds of hard rocks folded over into tight hairpin folds. There would seem to be no way a hard sandstone can be folded, yet the evidence that it has taken place is there in the rocks. Under certain conditions rocks can be deformed and the process is slow; were it to be rapid, the rock strata would fracture. In instances where successive layers of rock involve shales and limestones or sandstones, the thickness of the limestones or sandstones remains constant, but the shales are plastically deformed because they react to pressure more readily, becoming thin where they are stretched and thicker where they are squeezed together.

When a layer of rock is folded, the particles and minerals at the outer margin will be stretched and the rock thus weakened, whereas at the inner margin of the fold the particles will be forced together and the rock will be made more resistant. This phenomenon has a curious consequence because the parts that are folded upwards like an arch, 'anticlines', are the weakest and are thence more readily attacked by agents of erosion. However, the parts that have been downfolded in a trough, 'synclines', are more resistant, so as erosion proceeds, it is the downfolded parts that eventually become the higher ground.

When the forces that fold the rocks are such that the strength or brittleness of the rock will not permit further folding, the rock will fracture and one set of strata will be thrust over another in a reversed or thrust fault. Faults also occur where the forces have been either too strong or too rapid for the rocks to accommodate and they have simply parted asunder. If a force pulls apart the rocks and one side of the fracture slides down, this type of break is termed a normal fault. Some faults can be a major feature of the landscape. The Great Glen Fault, for instance, cuts right across Scotland and also forms the site of the inland lake, Loch Ness. When the rock structures on both sides of this fault line are examined, they do not match. However, if one side was to be shifted some 100 km alongside, the geological structures would match exactly. From this observation, it can be concluded that the northern part of Scotland has shifted some 100 km towards the north-east; from the age of the rocks, this took place during the Caledonian period of mountain-building 370 million years ago.

An examination of geological maps of most parts of the world will show rocks of different ages, represented by coloured areas, occurring in regularly repeated patterns because those particular strata have been folded. The folds have been levelled by erosion so that what one is looking at is a horizontal slice through the folds. Such folded strata always slope towards the younger rock. If a pile of sediments is folded up in a simple mound or anticline and this is then sliced horizontally, the material in the centre will represent the lowest stratum or oldest rock, whereas the original top layer of younger rock will be preserved as the outer ring. Wherever one examines folded sedimentary rocks, the slope or dip of the strata is from the centre of older rock down towards the younger sediments. It is because of this folding and subsequent erosion that geologists do not need to dig deeper and deeper to find older rocks and fossils. It is simply a question of walking up a slope to meet progressively older and older rocks which are outcropping at the surface.

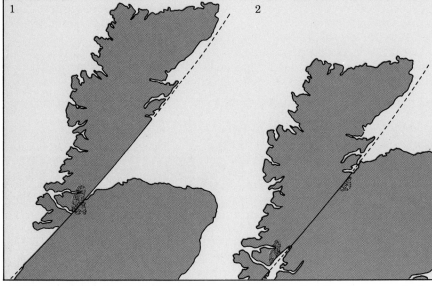

▲ 1 The Great Glen Fault, Scotland. 2 The geological structures on either side match, proving that they have moved a large distance relative to one another.

▲ The incised meander of the Wye, England is evidence that the land has been raised.

▲ The Lulworth Crumple, England, shows gentle folding.

▲ Major folding and faulting in rocks in Cornwall, England.

It is one thing to demonstrate conclusively that sedimentary rocks have been uplifted from the sea-bed to form dry land, but it is quite another to suggest what the actual mechanism may have been. James Hutton was faced with just this problem. He recognized that the weight of overlying sediment could have caused the underlying sediment to sink to great depths where heat and pressure caused it to consolidate into rock. He noted that granite must have been in a molten state to be forced up into existing rock and concluded that this had melted because of intense heat at depth. He believed that this was part of the process responsible for raising up the sea-bed to form dry land.

Hutton also realized that the piles of sediment that accumulated at the mouths of ancient rivers must have been thousands of metres deep since this was the thickness of the layers that could be measured in the rocks. It seemed evident, therefore, that the sea-floor must have sunk to accommodate the influx of fresh sediment, a process which must have been going on for hundreds and thousands of years.

While sediment was sinking, sea-levels were rising. In the mid-nineteenth century this was attributed to the shrinking of major ice-caps in the world which had caused the sea-levels to rise globally during the past 35,000 years. Evidence of sea-level rises is seen in the extinct coral reefs off the West African coast which are now at depths where corals cannot live. When the sea-level was lower, the rivers cut their valleys across what is now the continental shelf. Hence off the east coast of the United States, there are huge submarine canyons at the mouths of all the major rivers emptying into the Atlantic.

While rising sea-levels inundated the world's coastal areas, in Scandinavia the land was rising at an even greater rate. The Scandinavians could see that one-time ports had moved inland from the coast and were above sea-level. The land rose because Scandinavia was covered by a massive ice-cap, the weight of which caused the land to subside exactly as the sea-floor subsides under the additional weight of sediment. When the weight was removed, that is the ice-cap melted, the land sprang up again.

It was from this kind of observation that the idea of 'isostasy' (from the Greek words *isos* meaning 'equal' and *stasis*, 'setting') was put forward in 1889, by the American Clarence Edward Dutton (1841–1912). He suggested that the continents which are made of lighter material, the sial (silica and alumina), are 'floating' on a denser substratum of sima (silica and magnesium-iron material). The height of the Earth's surface is proportionate to the depth of its crust, so that if a section of the upper portion of the crust were to be removed, as in the erosion of mountains, the land would automatically rise up to compensate

The higher the land surface, the deeper is the crust; this means that mountains must have very deep roots. Evidence of this was discovered by the Frenchman Pierre Bouguet while surveying the Andes during an expedition to Ecuador in 1735. According to the law of gravitational attraction between bodies, a plumb-line would be deflected by the mass of a mountain range. However, Bouguet observed that it was actually deflected by less than had been calculated. This implied that the plumb-line was being attracted by a mass below the Earth's surface as well as the mountains above it and that therefore the mountains must have deep roots.

The principle of isostatic adjustment can be demonstrated by floating wooden blocks of different sizes

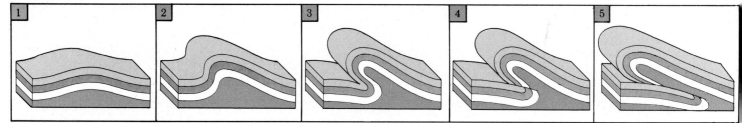

▲ **Rock folding**
1 Strata are lifted gradually. 2 With increasing pressure from one side the fold becomes assymetrical. 3 As the lateral pressure increases, the fold becomes tighter. 4 Eventually the rock fractures to form a fault. 5 A thrust fault develops with the upper part of the fold pushed over the lower.

▲ This landscape shows that the hills are composed of down-folded rocks, which are resistant to erosion. The valleys are formed by rocks which have folded upwards and are more subject to erosion.

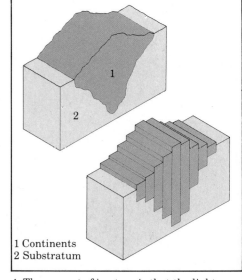

1 Continents
2 Substratum

▲ The concept of isostasy is that the lighter continents 'float' on a denser substratum. The blocks of wood floating in water demonstrate the principle, showing that if some of the upper surface is removed there is a compensating rise.

84-2885/MCL

in water. If a slice is removed from one of them, it will rise up so that it is floating with the same proportion of wood in the water as it had before. This important process causes vertical adjustments and explains some cases of uplift, like raised beaches. Unfortunately it does not seem possible to use isostasy to explain some horizontal or lateral movements of one portion of the crust relative to another over huge distances.

This phenomenon can be explained by other agents that are known to produce both vertical and lateral displacements of adjacent parts of the crust. During the voyage of HMS *Beagle* around the world (1831–6) Captain R. Fitzroy and Charles Darwin both recorded their experiences while they were in Concepción, on the Pacific coast of Chile, when a devastating earthquake coupled with extensive volcanic activity took place. The effects of this episode were felt over 1,000 km of coast. The city of Concepción was levelled and the port itself swept away by the giant tidal waves generated by the disturbance. Concepción Bay was raised permanently by a metre, while a nearby island was uplifted 3m. The association of earthquakes and faulting

in the lateral displacement of rocks is exemplified further by the famous earthquake of San Francisco in 1905 in which land was displaced 6m. The association of such events with volcanic activity indicates that deep weaknesses in the crust allow molten magma to reach the Earth's surface and presumably play some active and contributory role in uplift.

The area encircling the Pacific Ocean is frequently referred to as the 'ring of fire'. Earthquakes and volcanoes occur in this circum-Pacific belt, which is also noted for its mountains – starting with the Andes of South America, the mountain ranges go on into the Rockies of North America, extend to the Aleutian islands of Alaska, and continue in the islands of Japan, notorious for their earthquakes and also the volcano Fujiyama. There seems to be a definite connection between mountain chains, volcanic activity and earthquakes.

While this link seemed obvious to geologists, the process responsible for the connection was not known. There was a further problem for which no explanation was forthcoming. It had long been recognized from the data collected by geologists over many decades that the mountain-building

periods or orogenies were restricted to particular periods of time in the geological record. During the Devonian period, some 400 million years ago, there was a major mountain-building episode in eastern North America and western Europe which is called the Caledonian orogeny, in which the main mountain chains were aligned roughly north-north-east. In the same area 280 million years ago, there was the Hercynian or Amorican orogeny in which the main direction was more or less east-west, and then some 25 million years ago the Alpine orogeny occurred, stretching from Europe through to China, which produced the present-day Alps and the Himalayas. In western North America, Australia and the Antarctic, there were similar orogenic periods: some coincided in time but others, such as the Rockies, seem to have been formed about 70 million years ago in a quite different period of mountain-building. Although these facts of the geological record were firmly established, there was no explanation of why this should have happened until the theory of plate tectonics (see pp 178–91) was developed in the late 1960s. This overall synthesis helped towards a geological understanding of the Earth's history.

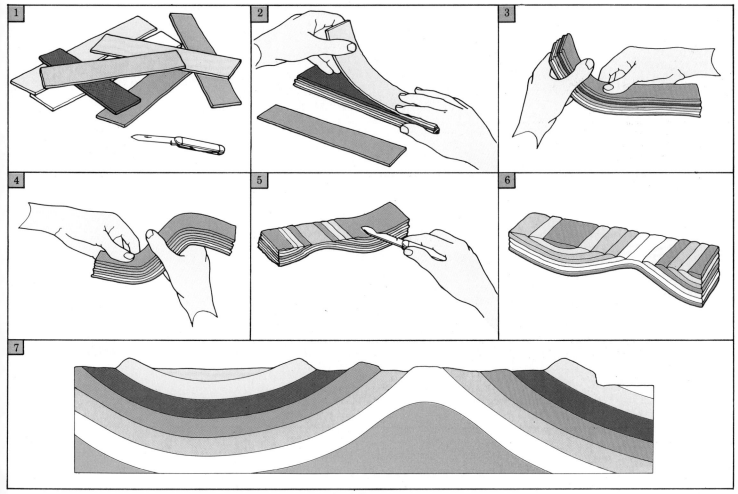

▲ By using layers of plasticine to represent folded rock strata and slicing them through horizontally, it can be shown that folding and erosion cause the different strata to outcrop at the surface.

LANDSCAPES OF ICE

**Clay containing rocks of all sizes was attributed to
the Flood until it was suggested that it was caused by glaciers.
Further evidence proved the existence of past ice ages.**

When we look at the solid rocks around us, like James Hutton 'we see no vestige of a beginning or prospect of an end.' Yet geologists could see there was evidence of rocks which had not been mentioned by Hutton. Over vast regions in Europe and North America, the rocks had a covering of clay mixed with pieces of rock of all shapes and sizes which were completely unsorted. Moreover, scattered over the landscape were huge boulders that had clearly been carried from distant places and ended up perched on the tops of hills and other unlikely spots. The clays that contained such a curious mixture of rocks were called boulder clay and are now termed 'till'.

This covering of clays and gravels, especially on the upper surfaces of hills, was interpreted by most scientists at the beginning of the nineteenth century as convincing evidence of the Biblical Flood – a sudden catastrophic event that left a layer of sediment plastered over the land once the waters had subsided. There was one problem: it was difficult to envisage how huge granite boulders could have been carried up onto mountain tops to be left stranded there. Although Lyell was obviously not a diluvialist, at one time he used the Flood as the basis of his Drift Theory to explain this phenomenon. He suggested that the boulders were incorporated in icebergs and during the Flood they drifted over what is now the land; when the icebergs melted, the boulders dropped onto the land beneath. It is from this theory that all the 'superficial' deposits (so-called because they are not of the same

geological structure as the underlying solid rock) came to be termed 'drift'.

It was not until 1838 that the explanation for these occurrences began to emerge. The Swiss geologist and world authority on fossil fishes, Louis Agassiz (1807–73), demonstrated that many of the features of the landscape could only have been produced through the action of ice. He was able to point to the work of glaciers in the Alps, showing how moving rivers of ice carved out U-shaped valleys with flat bottoms, as opposed to the V-shaped valley cut by water. Agassiz even went so far as to postulate the existence of great ice sheets covering most of Europe and

also North America. He was able to convince the leading geologists of Britain that many features of the landscape in Scotland and England proved the former existence of ice masses over the land.

One of the features of glaciers or ice sheets is that they pick up fragments of rock. These are carried along and are ground down to a rock flour as the ice moves over the land. They will also scratch the surfaces of boulders or rocks and such 'striations' are evidence of moving ice.

Where an ice sheet extends over a large area, rocks can be transported many hundreds of kilometres and when the ice melts, they will be left

▲ A glacial erratic in Yosemite National Park, USA. It is unrelated geologically to its site.

▲ Typical ice-carved scenery in the French Alps, with a glacier and sharply faceted mountains.

stranded far from their source. Such rocks are known as 'erratics' because they have wandered. Certain granite pebbles in the courtyards of some colleges at Cambridge University, England, were carried across the North Sea by ice from the granites of Norway to end up as pebbles in the gravels of Cambridgeshire. Where boulder-bearing ice reaches the sea, icebergs float away and when they melt, the rocks fall into the underlying sediments on the sea-bed. These so-called 'drop stones' deform the underlying sediment.

The edges of the glaciers or ice sheets are marked by huge masses of debris where all the material being carried is dumped. These structures are known as 'moraines'. The water from the melting of the ice will form streams and rivers, and coarse sands and gravels will be deposited in great delta-shaped fans. As they are the result of being washed out by the melt-waters these structures are known as

'outwash fans'. With the retreat of the ice, an extensive layer of boulder clay is left behind. As the ice retreats, the terminal moraines will be left in successive lines marking the various stages of the retreat of the ice. Often these moraines will block valleys so that the melt-waters become dammed to form lakes. This results in the formation of alternate thin layers of silts and clays; silts are deposited during the spring melt-water and clays during the rest of the year, giving rise to a series of fine laminations representing yearly depositions that are known as 'varves'. From these it is possible to count the number of years as with tree rings. Until the development of certain chemical techniques, counting varves was one of the main methods of dating material of the last ice age.

Till is characterized by containing pebbles of all sizes mixed with completely unsorted clays and sands. The individual pebbles or rock

fragments are scattered in the clays rather like currants in a cake and are not usually in direct contact with one another. When a till is exposed, the alignment of fragments with clearly defined long axes embedded in the clay can be measured with a protractor and a compass. This measuring of the orientation of the fragments can be plotted on a diagram to show the direction of flow of the ice sheet.

The other useful way of studying a till is to try to identify the different types of rock that make up the pebbles in it. If, for example, a till is composed exclusively of fragments of Chalk, then this suggests that the ice sheet has travelled over mainly Chalk country. In some regions the tills have carried huge blocks of Chalk, as can be seen along the north coast of Norfolk in England where the Chalk cliffs are surrounded by the clay of the till. Understanding the movement of clay and ice mixtures under ice-age conditions still presents many problems. For example, blocks of Chalk seem to have been squeezed and stretched almost like toothpaste. Even more curious are sands that probably were originally deposited in channels, which are rolled up in the till so that the sandstone appears in the forms of concentric rings.

Tills are clear evidence of ice ages and there have been four major periods of ice ages, lasting as long as 25 million years, in the Earth's history. As there are ice-caps over Antarctica and in Greenland, the Earth is considered to be still in an ice age. There are, in between the major advances of the ice, warmer interglacial periods and the one the Earth is experiencing now has lasted for 10,000 years.

The causes of ice ages, once it was agreed they had occurred, were a subject of considerable debate and it was not until the 1970s that the mystery was solved (see pp 166–7).

▲Varve clays at Uppsala, Sweden. The different bands represent the seasonal deposits of silts and clays.

▲Scratched surfaces of rock in Iceland caused by an ice sheet dragging rock fragments over them.

▲Modern till deposited recently on the Arctic island of Spitsbergen.

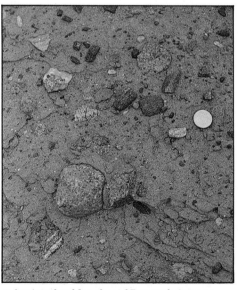

▲Ancient boulder clay of Precambrian age from Varanger Fiord, Norway

▲Lowestoft till from Suffolk, England deposited in the last ice age.

ICE MARGINS

Vast areas of land on the edges of the ice sheets go through cycles of freezing and thawing. This causes particular land features which provide further evidence of ice ages.

Evidence of past ice ages is easy to recognize from the way in which the landscape has been directly carved by ice and from the evidence of glacial deposits. However, further evidence is to be found in deposits around the margins of ice sheets, where there is a wide zone in which land is permanently frozen, a condition known as 'permafrost'. This land is the tundra where, during the summer months, the surface layers of the ground thaw and plant and animal life flourishes. Because these conditions exist around the margins of ice-caps, they are termed 'periglacial'. Intermittent

thawing and freezing causes numerous features to develop in the tundra which are capable of leaving a record in the rocks. Indeed, many features of the present-day landscape can only be understood in the light of past periglacial conditions.

One of the features of permafrost is patterned ground which is something akin to the cracking of mud into polygonal flakes as it dries out. As the water in the mud dries out, the surface shrinks and because it is unyielding it fractures. With permafrost, when the icy surface freezes, it shrinks and fractures in a polygonal pattern. This is because ice contracts when cooled further. Water collects in these fractures and later freezes in a solid block of ice. Such conditions are usually seen in cross-section and are known as 'ice wedges'. When the ice finally melts, the margins of the crack will often collapse, but usually sands and gravels are swept in, forming an ice-wedge cast. These structures occur in rocks such as gravels and sands where the bedding is suddenly interrupted by a wedge-shaped infill of

different sediment. Infills may continue over large distances of the surface and in some cases may join up to form polygons. The recognition of ice-wedge casts is crucial in elucidating the history of particular areas during ice ages. Even where there is no sign whatsoever of the action of glaciers, ice-wedge casts are evidence of permafrost and tundra and, therefore, a glacial episode.

A feature of modern tundra conditions is the presence of small perfectly rounded hills standing proud above an otherwise flat and monotonous landscape. These hills are known as 'pingos' and are again due to disturbances resulting from the freezing of water. Pingos develop either in old lake basins or are fed by groundwater. When water freezes it expands and pushes the ground upwards. In a fairly homogeneous sediment, this heaving produces circular mounds. In fact, the process of pushing up the ground creates more space, more ice forms and so the pingos grow. When the ice eventually melts, these landforms collapse in the middle

▲ Aerial view of glaciers in Greenland.

▲ Modern ice wedge filling a crack in sands in the Canadian Arctic.

▲ Fossil pingos in Norfolk, England are evidence of past tundra conditions.

to form circular rims usually with a central lake or boggy ground. The formation of pingos over a period results in newer examples cutting into pre-existing ones, so that when the entire family of pingos have collapsed there will be a pattern of intersecting rings. This produces a curious land surface comprising circular ramparts several metres high, often intersecting one another, with a flat boggy central area. When viewed from the air, this type of periglacial landscape can be readily recognized as having been made by pingos.

The action of freezing and thawing in the ground is seen in sections of ancient soils and subsoils which look as though they have been violently churned up. At first glance their appearance as a mixture of different rock types is similar to that of a flash flood deposit, but this impression is misleading. As water freezes, rock fragments are pushed up and, when the water thaws, they fall back; but when this process is continually repeated, the soil becomes churned and the rock types are mixed

up. This is given the name 'cryoturbation', from the Greek *kryos*, frost. In gravels affected by this process, the flat pebbles tend to end up vertically which cannot happen during the normal processes of deposition. Any gravel showing this type of fabric is evidence of tundra conditions. that existed in the past.

Frost shattering of rocks that were exposed at the surface is a further indication of the previous existence of periglacial conditions. The clearest example of this is seen wherever Chalk outcrops, either on a sea cliff or in a quarry face, and the uppermost surface is broken into angular fragments to a depth of several metres. As Chalk is exceptionally porous, water soaks into it, and when it freezes and expands, the Chalk is broken up. This cracked Chalk is termed 'head'. The evidence of the first ice age to affect Britain was based only on the appearance of this cracked Chalk.

Chalk itself provides some of the most surprising evidence of permafrost of the periglacial regions. Chalk gives rise to soft rolling countryside in which

there are numerous dry valleys; there are no rivers or streams in them, because water simply soaks through Chalk and does not easily flow over its surface. Yet it seems clear that the valleys were carved out by water flowing over the landscape. This could only have occurred at a time when water was unable to soak through the rock. The only circumstance that can be envisaged is that the region was one of permafrost; any flowing water would have been on the surface and hence capable of eroding the landscape. The rolling downlands of the Chalk countryside, as in the Sussex South Downs, England, provide eloquent evidence of the former ice age.

Further evidence of ice ages is found in the occurrence of wind-blown silt or 'loess', which can accumulate tens of metres of thickness. Strong winds develop around the margins of ice sheets or cold deserts, such as the Gobi Desert, and can transport fine materials hundreds of kilometres. This fine yellow loess, sometimes known as 'brickearth', is again characteristic of periglacial conditions.

▲ Ice-wedge cast Norfolk, England. The darker sediment has filled in a fracture, which contained an ice wedge.

▶ A modern pingo in the Canadian Arctic which has formed because the frozen underlying sediments expand and push upwards.

▲ These disturbed sediments in Norfolk, England are the result of alternate freezing and thawing in tundra conditions.

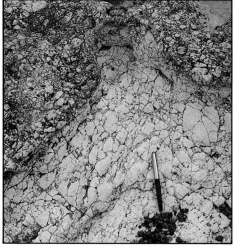

▲ Fractured Chalk from East Anglia, which is the only evidence of the first ice age in Britain.

CHAPTER 2

HOW TO UNDERSTAND FOSSIL EVIDENCE

Fossils are the remains of past life.
It is not always the living things themselves which
are fossilized, but sometimes merely the traces
of their former existence, such as the impression
of a shell in a rock or a footprint on an ancient mud-flat.
The real problem in evaluating the fossil record
is that much of the evidence is missing, because the
conditions for preservation were not always favourable.
Only certain organisms stand much chance of
being fossilized, such as those that have a mineralized
internal or external skeleton. Even this may not be enough;
preservation also depends on what kind of
environment the organism inhabits.

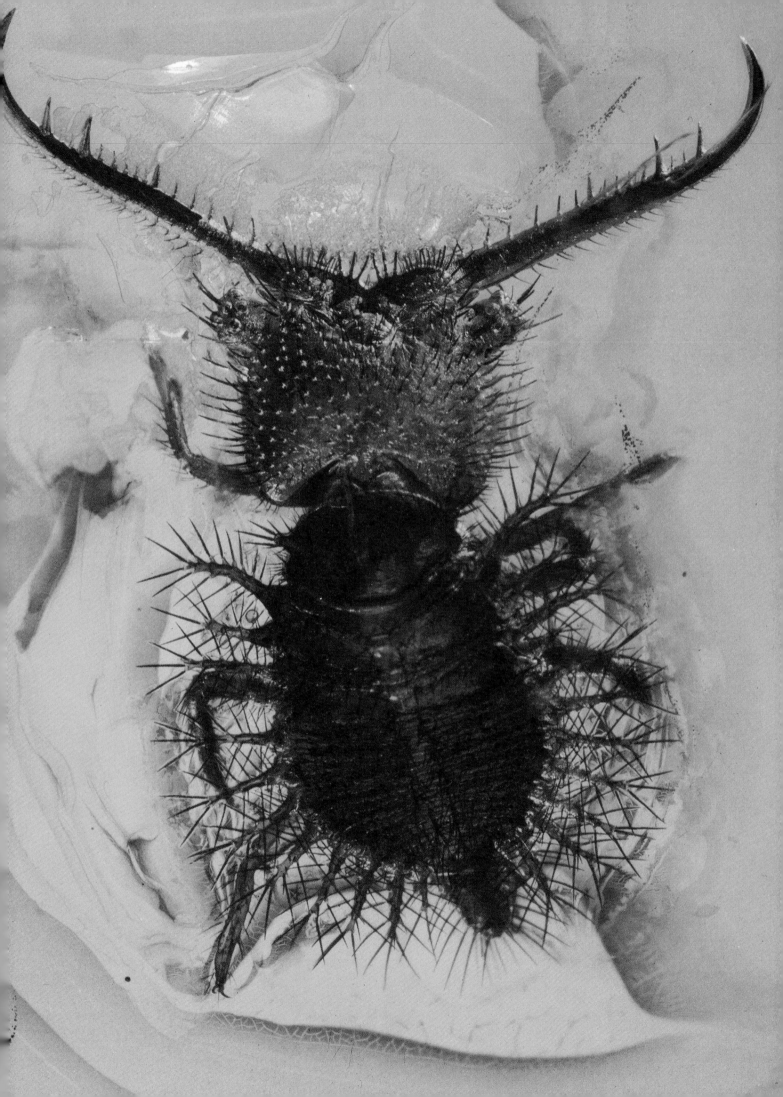

THE PROCESS OF FOSSILIZATION

Most fossil shells and bones have a stone-like quality. This is because of changes in their mineral composition.

In the eighteenth century, collectors called any object dug up out of the ground, including minerals and crystals, a 'fossil' from the Latin *fossilis*, something dug up. By the beginning of the nineteenth century, the word fossil had acquired its present-day meaning of a relic or trace of past life preserved in the rocks.

A fossil can be in the form of a preserved hard part of an organism, such as a shell, bone or tooth, or can be a trace made by the animal when it was alive, like a tunnel through sediment or a boring into a piece of rock. Some rocks are composed entirely of fossils, for example limestones which consist mainly of fossil shells. Oil and coal derived from the decay of animals and plants over a very long timespan are known as fossil fuels. But once the decay has progressed to a stage where the original material is utterly

unidentifiable, then it is no longer justifiable to use the term fossil.

One question arises: at what point do we start calling preserved remains, fossils? This is difficult to decide and in fact has been settled by inventing an arbitrary rule. The agreed convention among scientists is that remains which have been preserved in rocks laid down since the last ice age, that is, younger than 10,000 years, are not usually classified as fossils (although sometimes they are called subfossils). There is another approach whereby remains are described as fossils: if there has been an obvious change in the structure of the bone or shell, such as petrification. This has its pitfalls since fossils tens of millions of years old may have suffered no discernible changes. In practice it is not difficult to recognize whether a specimen is a genuine fossil or not, and

the real task is to work out how it was preserved and what has happened since the animal or plant died.

When an animal dies the first consequence is that the soft parts are decomposed mainly by the activities of fungi and bacteria. If the material is preserved in conditions where normal processes of rotting do not take place, then the organic matter will be slowly converted to a carbon film. This is most commonly seen with fossil plants, like leaves which leave their imprint on rocks. Plant tissues are composed of carbon, hydrogen and oxygen, which make up carbohydrates such as sugar, starch and cellulose. Plant material breaks down much more slowly when it has been buried; because oxygen is unable to reach it, the oxygen and hydrogen are lost and the carbon is left behind. The carbon is capable of further combustion, but if the process

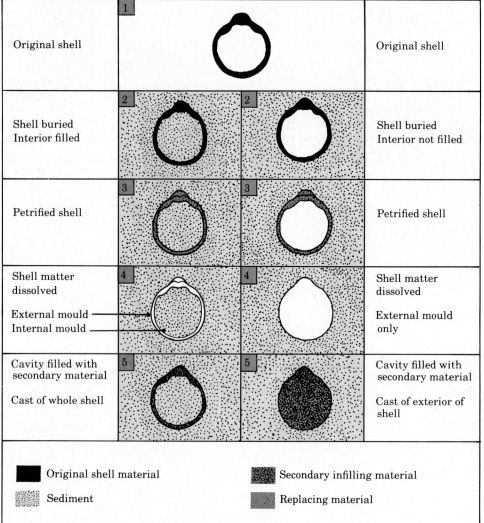

Original shell	1		Original shell
Shell buried Interior filled	2	2	Shell buried Interior not filled
Petrified shell	3	3	Petrified shell
Shell matter dissolved / External mould / Internal mould	4	4	Shell matter dissolved / External mould only
Cavity filled with secondary material / Cast of whole shell	5	5	Cavity filled with secondary material / Cast of exterior of shell

■ Original shell material   ▒ Secondary infilling material
░ Sediment   ▓ Replacing material

▲ Two methods in which a cockle shell can become a fossil.

▲ Fossil of the exterior of the shell *Trigonia*. Internal cast of *Trigonia*.

is halted a fossil will result.

With shells and bones, the changes may be less obvious. The tiny spaces in the bone or shell can become impregnated with minerals deposited from percolating waters, such as calcium carbonate. Often the nature of the original mineral is altered. The mineral in bones and teeth is calcium phosphate. In fossil bones, the element fluorine, which is present in minute amounts in most water, becomes incorporated into the calcium phosphate crystal; this makes the bone mineral more resistant to erosion.

These changes that occur in fossils and which make them more stone-like are known as petrifaction, which means 'made into stone'. In some instances wood, as in petrified forests, can be replaced by such minerals as silica; this commonly occurs molecule by molecule so that sections of the silicified fossil wood, cut on a diamond wheel, reveal microscopic details of the wood tissue. This replacement is known to have happened with the skeletons of 500 million-year-old fossils, the graptolites, which look like writing on the shales

in which they were preserved, but were in fact small colonial marine organisms distantly related to the vertebrates. In these tiny skeletons the fibres of the protein collagen are preserved and can be studied under the electron microscope. No original protein survives: instead its exact configuration is preserved by the mineral.

In Australia some fossils are transformed into semi-precious opal, a type of silica which is not completely crystallized; it contains very fine crystals which scatter light to produce the 'fire' in the stone. Occasionally entire bones of prehistoric reptiles, such as plesiosaurs and dinosaurs, can be preserved as large pieces of opal. The original mineral of the fossil can also be recrystallized, meaning that further crystals are added on to the original crystal structure. This has happened, for example, in the spines of sea urchins and parts of other echinoderms. In life the limy skeleton would have a spongy texture; but in the fossil, calcium carbonate deposits fill in the holes to form a single large solid crystal. Nevertheless, these changes do not appreciably alter the

shape of the original skeletal object.

When the articulated valves of shellfish are buried, several events can take place. If once the sediment round the outside of the shell has consolidated, the actual shell or bone may dissolve away to leave a cavity. Many shell beds are of this nature; the sediment, often sandy ironstones, is a mass of external moulds. Sometimes the hollow shape is later filled in by another mineral so that a cast of the original shell is made. It was this kind of preservation that led scientists in the seventeenth and eighteenth centuries to believe that these shells of rock were formed in the ground and were not truly the remains of shells.

This process can be taken a step further. Where a fossil has a natural cavity, as inside two valves, this space may be filled with sediment or mineral deposits; if the fossil itself is dissolved away, the resultant fossil made of the sediment will be a cast of the internal mould of the original fossil. This commonly occurs with ammonites, where the chambers of the shell originally filled with gas are filled with sediment and the shell then dissolves.

▲ Part of a shell bank of brachiopods (Palaeozoic shellfish) which shows that both original shells and moulds have been preserved.

▲ Ammonite fragments in which the separate gas chambers have been infilled by calcium carbonate. The original shell has dissolved away and the infillings interlock like a jigsaw.

▲ A fossil which has been split open to show that the complete ammonite together with an impression of the shell have been preserved.

FOSSIL ENVIRONMENTS

Most familiar animals and plants have little chance of becoming fossils. There are very few places on Earth where the right conditions exist. This makes the fossil record unbalanced.

Living things inhabit most parts of the surface of the Earth, from the tops of mountains down to the deepest depths of the oceans, yet their chances of becoming preserved are severely limited. By far the most important prerequisite for becoming a fossil is rapid burial. This automatically restricts potential fossil-bearing areas to those where deposition is taking place. On dry land, where most animals and plants are observed, virtually no fossils are formed.

When it comes to considering life in the oceans, there is a problem of a different kind. Microscopic skeletons of animals and plants inhabiting the world's oceans are continually accumulating on the floors of the oceans. However, while 75% of the land surface consists of sedimentary rocks, virtually none come from the deep seas. This means the oceans do not contribute much to the fossil record. It might be imagined that drilling deep into the sea-beds would give a continuous record of oceanic life through geological time. One of the great mysteries that faced geologists when they began to obtain evidence from the modern oceans was that there did not seem to be any rocks much older than 100 million years. The ocean floors, seemed geologically to be surprisingly young (see pp 186–7).

The greater part of the land and the ocean basins for most of geological time seems to be singularly unrepresented in the fossil record. It is the narrow coastal zones of land and sea where animal and plant remains stand the greatest likelihood of being preserved. This is because sediments derived from the erosion of the land are transported to the sea; as they reach the coast, sediments are dumped and gradually build up to form flat alluvial plains. The shallow seas around the margins of the land also receive sediments and nutrients. As the sea is shallow, sunlight can penetrate and plant life is able to flourish on the sea-bed. Similarly, large numbers of animals inhabit the sediment, feeding on the plants and on one another.

The peculiar bias of the fossil record is that most living things observed in everyday life have little chance of ending up as fossils, whereas the forms of life that live in the sea, on and in the sediments of the continental shelves, make up the majority of fossils. The near-shore animals and plants with mineralized skeletons will dominate the fossil record, as will the

▲ Fossils are unlikely to be preserved in upland areas because this is where erosion takes place.

▲ In deserts animal remains are likely to be buried by sand and fossilized.

▲ In shallow seas where sediment is being deposited, skeletons can be buried.

▲ **The environments in which fossils are found.**
Fossils are restricted to a comparatively narrow zone of the Earth's environments – to areas of deposition. On land these are inland lakes, deserts, alluvial flood plains and deltas. In the sea it is only in shallow water that fossils are normally found.

shelled animals living within the sediments. Land animals that inhabit flat alluvial plains have a chance of becoming fossils, especially if they fall into a river, get stranded on a bank and covered by a sand shoal, or get swept downstream.

If one observes conditions both on land and at the seaside, it is difficult to envisage how any animal remains are ever preserved. It is at the very times when people are not likely to be on the beach or wandering over the countryside that conditions are favourable for fossilization. Storms that lash the coast, causing rivers to overflow and flood lowlands, are times when animals are caught up, swept to their deaths and buried by sands and muds. Marine organisms living within the sediments are likely to be caught up in storms and dead shells will be winnowed out of shell banks.

Other environments which provide conditions in which organisms are frequently preserved are lakes and swamps. Lakes are shortlived in geological terms because as sediments are brought in by rivers and streams, they will eventually become filled up; if

there is no renewal of water supply, then the lake will dry out. The most sensational evaporation of a lake in geological history was that of the Mediterranean – known as the Miocene salinity crisis of the Mediterranean. It was one of the largest lakes ever formed. Africa collided with Europe at the Iberian peninsula and Asia in the Middle East cutting off part of the sea to form a lake. This enormous lake in fact dried out leaving huge salt deposits where nothing could survive. The Mediterranean refilled from the Atlantic over the Straits of Gibraltar about 7 million years ago.

Lakes accumulate sediments, and snails and bivalves, in particular, are preserved in abundance. In Lake Turkana, Kenya, a detailed record of the evolution of snails during a continuous period of 5 million years was reported in 1981, in one of the classic documented studies of evolution in an isolated restricted environment

A major terrestrial environment in which vertebrate remains may be preserved are desert regions. Where sand is being transported by wind,

there is the frequent chance of a sudden sandstorm engulfing animals. If these are not immediately consumed by scavengers they may become mummified so that even the texture of skin, as in many dinosaur remains, may be perfectly preserved as imprints in the sands which covered them.

In the seas, away from sources of sedimentation, there live animals that secrete skeletons of calcium carbonate and build up large reefs. These in turn become inhabited by numerous other organisms, some encrusting the reefs, others seeking shelter and even boring into them. As these can only exist in shallow waters, they tend to be on the continental shelves, although they also form around oceanic islands.

There is another major environment in which animals and plants may be preserved and that is in caves (see pp 54–5). Although caves develop in upland areas, if the seas flood over the areas faster than the rate of erosion, then cave systems may be buried. The remains within them will be preserved until the ancient landscape is exhumed following the next cycle of uplift and erosion.

▲ Wildebeeste which have drowned in a river. Many fossils are found in river deposits.

UNUSUAL FOSSILS

Some seemingly insignificant holes in rock have been found to be valuable fossil evidence. Other rare fossils can preserve complete animals, proteins, blood cells and even colours.

Important fossils can very occasionally be preserved in such a way that they may be overlooked for many years and their significance go quite unrecognized. In a quarry at Daleszyce in the Holy Cross Mountains of central Poland, a quartzite of Lower Devonian age (410 million years old) was known to have remains of primitive armoured fishes, but their fragmentary nature led geologists to believe that they were of little scientific value. The sandstone in the quarry was formed from rounded grains that had been cemented by silica which had overgrown the original grains. The special feature of the fossils which caused them to be ignored was that they were preserved as holes in the rock. The bony materials had dissolved away to leave perfect moulds in the rock. The study of this material, which began in 1955, involved first collecting all the pieces of rock in which there were obvious hollows or impressions. The material was then taken back to the Polish Academy of Sciences, Warsaw, and rubber latex casts were made of all these natural moulds. First the surfaces were dampened with ammonia, as this ensured that the latex found its way into the smallest indentations, and then a plasticine dam was built around each specimen and the latex was poured in. When the latex was set and peeled off, it produced a perfect replica of the original surface of the bony armour of the primitive fishes. The fine details of the ornamentation, as well as the individual bones of the armour, could

then be examined; the rock acted as a kind of master mould. Among the armoured jawless fishes that were discovered by using this method were six new species, the best known being *Guerichosteus kozlowskii* and *Hariosteus kielanae*, which were easily identified by their ornamentation. One specimen of a hitherto unknown *Rhinopteraspis* revealed new details of the anatomy of the mouth region of this group of fishes. The holes in this Polish quarry have proved to be an important source of fossil evidence and there are still many new types of jawed fishes from it to be studied.

Another famous deposit in which the only evidence of fossils was a series

of holes in a wall of a sandstone quarry comes from the Upper Triassic (200 millions years old) Elgin Sandstone from north-east Scotland. In 1964 Dr Alick Walker of Newcastle University, England was able to reconstruct the complete skeleton of the primitive bipedal carnivorous dinosaur *Ornithosuchus* from holes in the rock. He filled the holes with a flexible plastic, which when removed gave a replica of the detailed shape of the individual bones from which he eventually managed to reconstruct the skeleton. The casual visitor to one of the quarries around Elgin would probably have ignored the occasional round hole in the quarry face, not

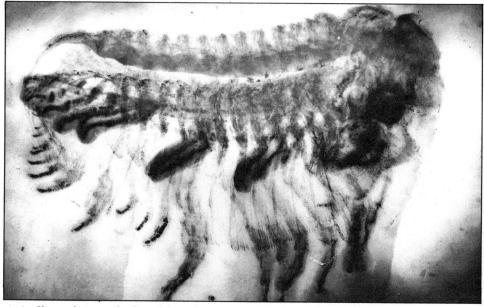

▲ An X-ray photograph of a trilobite showing the animal in great detail – the legs and compound eyes can be clearly seen.

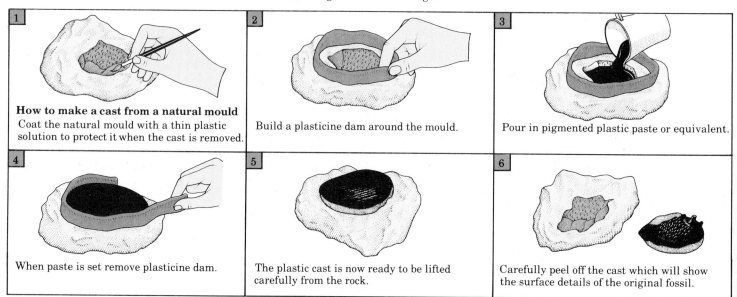

1

How to make a cast from a natural mould
Coat the natural mould with a thin plastic solution to protect it when the cast is removed.

2

Build a plasticine dam around the mould.

3

Pour in pigmented plastic paste or equivalent.

4

When paste is set remove plasticine dam.

5

The plastic cast is now ready to be lifted carefully from the rock.

6

Carefully peel off the cast which will show the surface details of the original fossil.

suspecting for one moment that he was looking at the cross-section of where a dinosaur limb-bone had once previously been.

The preservation in the Roman city of Pompeii of objects, including people, overwhelmed by the eruption of Vesuvius in AD 79, was similarly in the form of holes, although because of their date, these remains cannot be classified as fossils. The volcanic ash engulfed everything and when it solidified it left a perfect mould of the buried objects. This was because the organic remains completely disintegrated and vanished, leaving a hollow space in the solid ash. When plaster was poured into the empty hollows human bodies, animals and food were revealed in the form of plaster casts.

At the other extreme of preservation there are fossils that appear to be virtually unchanged. Among the most remarkable discoveries of recent times was a complete frozen baby mammoth in Siberia, found in 1977. The red and white blood cells are intact and so are the proteins of the tissues. It is possible to work out the sequence of amino acids in the blood respiratory pigment, haemoglobin, and to compare it with the sequences in both the Indian and African elephants, which establishes the genetic relationship of the woolly mammoth to the living elephants.

One technique for studying fossils which throws an unexpected light on some common fossils such as the trilobites are X-rays, which have revealed details of limbs, alimentary canal and eyes which would otherwise be unknown. Another technique is to make microscopic sections of shells, bones or plant material which can provide clues about different aspects of the mode of life of the animals and plants concerned. Twenty years ago it was discovered that the original protein, collagen, of fossil bone was capable of being preserved. This can be examined under the electron microscope and the characteristic appearance of collagen banding can be seen. This observation has led to a whole research programme extracting and measuring the exact amounts of different amino acids in fossils. The most recent development in the scientific study of fossil proteins is that the intact protein chain can be extracted and split into specific segments using enzymes. The existence of fossil proteins has led to a new branch of palaeontology – palaeobiochemistry – and it is hoped eventually to discover fossil evidence of evolution at the molecular level.

Finally there are a number of quite bizarre types of preservation which we cannot as yet explain. One example from 25 million-year-old Miocene rocks at Lake Rusinga, Kenya, is a truly remarkable fauna as yet not fully documented. There is a fauna of soft-bodied animals preserved as solid objects, including caterpillars preserved in the round; animals with skeletons are also preserved. Lizards and birds are fossilized complete with their muscles, which are transformed into calcium carbonate. One lizard can be identified as belonging to a present-day group because its forked tongue is preserved.

Some examples, such as the brilliant yellow-bodied flies from the 26 million-year-old Oligocene Insect Bed in the Isle of Wight, England, are due to a yellow pigment which is chemically inert. Frequently fossils such as the Eocene (38 million years old) snail *Theodoxus* turn up with the original colours and patterns still intact. For prehistoric land reptiles, such as dinosaurs, there is no direct evidence of their colours. However, the ocean-going reptiles, the ichthyosaurs, sometimes have their skin and pigment cells preserved showing that their colour was tortoise-shell.

▲ In 1977 a baby mammoth was discovered in Siberia frozen in the ice. It was so deeply frozen that the skin and soft tissues were preserved.

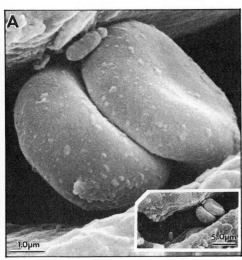

▲ Even the blood cells of the mammoth were preserved. Here are some magnified red corpuscles in position in a thin blood vessel.

▲ Fossil chameleon showing head and neck with bones and skin preserved in a Miocene deposit from Rusinga Island, Kenya.

THE FORMATION OF DEATH AND LIFE ASSEMBLAGES

Organisms are usually preserved away from where they used to live, in a place protected from erosion; this is termed a death assemblage. Those preserved in their habitat are known as a life assemblage.

Most animals and plants are not likely to become fossilized in their habitats. In fact the majority of living things are not preserved as fossils at all simply because they live on dry land which is subject to erosion. A fossil record is only produced under certain conditions.

When an animal dies it is first attacked by scavengers, then insects, and finally by fungi and bacteria, which utilize all the available nutrients. Eventually, all that remains are the bones and teeth, but even these are destroyed in time, the teeth surviving the longest. In order to stand any real chance of becoming a fossil, an animal or plant must be transported to a suitable environment in which sediment is accumulating.

When an animal falls into a river and is swept downstream, there is a very good chance it will be preserved as a fossil. While the corpse is carried along, gases form as it rots, helping it to float. After some time the carcass disintegrates and the skeleton falls to pieces so that the bones sink down and become mixed with the gravel on the river bed. Articulated pieces of skeleton may be stranded on sand banks and then buried by channel sands. Ancient river gravels frequently contain the bones of large animals which are often water-worn as a result of being dragged along the river-bed together with pebbles and stones.

As particular conditions are necessary for the formation of fossils, the likelihood of obtaining a representative sample of past life in a given area is very small. This can be demonstrated by choosing a small area of ground, about 1m^2, and making a list of all the animals and plant material found there, such as slugs, snails, woodlice, earthworms, millipedes, leaves and roots. Insects like snails and woodlice have hard parts capable of preservation, and hence are potential fossils. Now take a sample of this same area, say 100 mm^2,

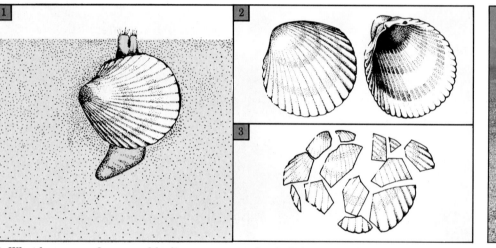

▲ **What happens when a cockle dies**
1 The life position 2 When the cockle dies the elastic ligament springs the valves apart and as it disintegrates the valves separate.
3 The valves become fragmented.

▲ A chenier is a shell bank accumulated as the sea sweeps up the shells on the strand line and winnows them out of the sediment.

▲ The Pliocene Red Crag in Suffolk, England is a shelly sandstone showing the accumulation of separated shells, convex side up, in a giant submarine dune.

◄ The strand line is marked by broken and scattered shells moved from their life positions by wave action.

dig the soil to a depth of 100 mm and sort out how much evidence there is of the remains of the animals already listed. This will give the overwhelming impression that the fossil record is very poor indeed. Yet everyday experience shows that this is not so. Many sedimentary rocks are full of fossils, and some are composed entirely of them.

The chances of organisms in shallow seas becoming fossilized are greater than in any other environment. A common bivalve such as the cockle *Cardium* provides a good example of how an underwater organism becomes fossilized. This animal lives buried in sand, filtering its food from the water by means of a short siphon that protrudes above the bottom sediment. Normally there is no sign of these shellfish alive because they are part of the in-fauna, that is, they live within the bottom sediment. When they die their soft parts are eaten by scavengers

or simply rot away; the two valves spring apart as the muscles which held them together perish.

The dead cockle lies beneath the surface of the sand. During a storm, when the sediment is churned up by the waves, the saucer-shaped shells become separated from the sand and are swept towards the shore. In this way cockle shells are scattered over the sandy beach. They have been removed from their original life position and lie with their valves either open or completely separated with the convex side upwards. If the valves land concave side upwards like a saucer, any slight current will turn them over. Once reversed, it becomes more difficult for currents to raise the valves, so this position is the more stable one.

As time goes by, the tides tend to sweep up the shells so that they accumulate at the strand line. Where the sea is slowly advancing landwards, due either to the land subsiding or the

sea-level rising, there will be an increasing accumulation of shells which may form an enormous bank. Every cockle in the vicinity will stand an excellent chance of ending up in the shell bank. Where such shell banks are preserved, like the examples in the Mississippi delta, the fossil record is remarkably complete.

This type of accumulation, which can be observed along most sandy coasts all over the world, is known as a death assemblage as the association of the shells was formed after the death of the animals. In no way does it reflect their association during life. However, if there was a sudden influx over the sediment or a sudden retreat of the sea and all the cockles were killed in their normal habitat, they would be preserved in their life positions. This is termed a life assemblage because it preserves the conditions and associations that existed when the animals were alive.

▲ A fossil death assemblage of complete and broken mollusc shells.

INTERPRETING FOSSILS IN THE ROCKS

**When fossils are found preserved in rocks
it has to be determined whether they represent a life or death
assemblage, which is not always easy.**

In the case of rather inactive organisms like bivalves, there is little difficulty in deciding whether they have been transported from their life position. But with highly mobile animals it becomes something of a problem to decide whether the remains should be considered as life or death assemblages.

Where animals are preserved in the position in which they died, it seems logical to consider the remains as a 'life' assemblage, even though it actually records the last few moments of their life. The famous collection of the armoured herbivorous semi-aquatic reptile *Aetosaurus* is an example of this, but it is reasonable to assume that these animals were not habitually in such concentrations. There are several possible explanations: perhaps

they congregated in the same pool for the mating season and the entire group was overwhelmed; or perhaps during a drought they retreated to the last pool, which finally dried up. Whatever the cause of their death, it is evident that the carcases remained undisturbed and were not even dismembered by scavengers. If we use the same criteria which we applied to the bivalves (pages 44–45), then this set of remains should be designated a life assemblage. In practice, if skeletons are preserved reasonably complete, it is assumed that they are life assemblages unless there is definite evidence to the contrary. Among the most famous dinosaur skeletons, again preserved in their death position, are two perfect examples of the early carnivorous *Coelophysis*. From comparing this slab

(page 47), with the aetosaurs, it would seem reasonable to conclude that these two skeletons represent a pair of dinosaurs that were preserved exactly where they had died. Unfortunately such a conclusion would be entirely mistaken because these skeletons are merely two out of dozens preserved on the same slab. Owing to the profusion of bones all piled on one another it is virtually impossible to discern the

▼A string of vertebrae found in 1909 by Earl Douglass was the key to a treasure trove of fossils in the Carnegie Quarry, Utah. This bone map shows that part of the quarry face is now an exhibit within the Dinosaur National Monument museum. The other fossils have been removed to other museums throughout America.

▶Perfectly preserved skeletons of the armoured, herbivorous, semi-aquatic reptile *Aetosaurus*. This specimen is 200 million years old and is now in the Stuttgart Museum.

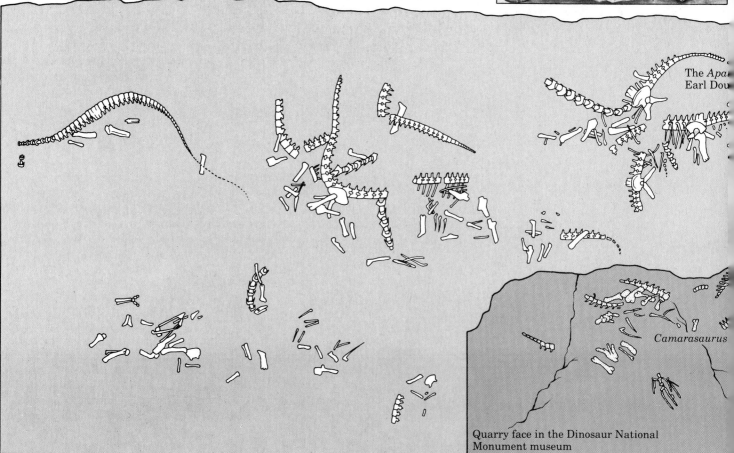

The *Apa*
Earl Dou

Camarasaurus

Quarry face in the Dinosaur National Monument museum

separate individuals. Therefore all but two have been carefully covered over for the sake of clarity. This pack of animals ranges from 1 to 3 m in length and some adults have young preserved within their stomachs, showing unique evidence of cannibalism in dinosaurs. There can be little doubt that the *Coelophysis* pack met its end together, but exactly how remains a mystery. But what is equally evident is that the site in New Mexico, from which their remains were excavated, is not the place in which they died. The dead or dying bodies were swept into this small area and came to rest in a heap. Again using the criteria applied to the bivalves this group is designated a death assemblage.

A now classic example of a complete fossil skeleton is the first bat in the fossil record, which comes from the bed of a 50 million-year-old Eocene freshwater lake. The bat's perfect preservation suggests that it was fossilized not far from its normal habitat. No one would imagine, however, that this bat lived on the bottom of a lake. But a reasonable inference to be drawn from the fossil is that an insect-eating bat flying over the waters of the lake, fell in, drowned and eventually sank to the bottom to be covered in lake muds.

The preservation of dispersed and disarticulated skeletons is usually interpreted as fairly conclusive evidence of the animal remains having been transported from their original habitat. But even here it is not always as simple as it might seem, as was proved by some observations made in East Africa. During a drought there in 1973–74 in which the elephant population was reduced by 75 per cent, the disintegration of the carcases was monitored over a number of years. There were too many to be dealt with

by vultures and hyenas so most were devoured by invertebrates – fly maggots and other insects and bacteria. The wet stage of decomposition lasted only two weeks; a further three weeks allowed *Dermestes* beetles to remove all the skin and sinews. Within five weeks of death the difference between the night and day temperatures (20°C) resulted in the bones cracking and flaking. The skeletons lasted longer in the dry season because the skin remained intact for longer. But even so it was not more than three to five months before the bones and teeth began to crack and flake. After a year the skeletons were completely disarticulated and separated. After two years only a few parts of the skeleton survived and were partially covered by soil; these bones alone were candidates for possible preservation as fossils. Such remains if found in rocks would unquestionably be interpreted as a death assemblage and it would be assumed that they were the result of dispersal and transport. Yet the bones have, in fact, hardly moved from the

▲ A photograph by Earl Douglass in 1910 shows the removal of an *Apatosaurus (Brontosaurus)* thigh bone from the Carnegie Quarry.

▲ Two complete skeletons of *Coelophysis*, 200 million years old, from the Upper Triassic period. These are now in the American Museum of Natural History.

...e first spotted by

Camarasaurus

One of the most nearly perfect dinosaur skeletons ever found: *Camarasaurus*, now in the Carnegie Museum, Pittsburgh.

...*gosaurus* plates

Allosaurus

Fossil tree trunk

Apatosaurus bones

position in which the animal died.

Among the most famous deposits of fossil bones is the outcrop of the Upper Jurassic Morrison Formation in Utah, USA, which was discovered by Earl Douglass, an indefatigable collector from the Carnegie Museum of Pittsburgh, in 1909. Out of the Carnegie Quarry came such famous dinosaurs as *Stegosaurus*, *Apatosaurus* (*Brontosaurus*) and the flesheater *Allosaurus*, together with the complete skeleton of *Diplodocus*. From 1909 the Carnegie Museum excavated dinosaurs and in 1915 this quarry and much of the surrounding countryside was created a National Park by the American Government. It was not until the 1950s that a building was built over the rock face to establish a museum on the site showing the dinosaur bones carefully exposed from the rock, but remaining in the position in which they had been deposited. This museum is unique in that it demonstrates the exact manner in which the bones are preserved. The bones are undoubtedly a death assemblage, the result of carcases,

parts of skeletons and dissociated bones being swept down a river to become dumped on a sand bank in a kind of log-jam of bones and bodies.

The famous deposit of Dura Den in Scotland is a small, saucer-shaped area of sandstone packed with fossils of complete fish. These fish inhabited flowing rivers and freshwater lakes throughout the northern hemisphere, but like the aetosaurs never lived in such profusion. In this example several different kinds of fish were trapped in a pond that had become cut off from the rest of the river system, perhaps during a drought. Eventually the pond too dried out and the entire fauna was preserved after rapid burial by the sudden incursion of sediment. This unique deposit is a life assemblage because it shows the conditions in which the fish lived immediately before the sediment covered them.

In marked contrast are petrified tree trunks which are sometimes found preserved exactly where they were once growing. However, frequently the trunks will rot, fall down and be washed to the sea; many so-called petrified forests are logs that have been

washed onto a bank during their journey downstream. But there is rarely any problem in deciding whether or not a fossil tree trunk has moved from its original environment, particularly if the trunk is still standing with its root system intact. If the trunk is lying horizontally in the sediment it clearly has fallen and moved and the only problem is to find out how far it has moved or been transported. Many specimens of fossil wood show borings made by the wood-boring bivalve *Teredo*, commonly known as the ship-worm. Whenever such examples are found the pieces of wood have been washed out to sea where *Teredo* lives. The evidence of *Teredo*-bored wood is of a life assemblage as far as the bivalve is concerned, but of a death assemblage with regard to the original tree.

It is the same situation with animals that live in association with others. Organisms such as oysters, serpulid worms, bryozoans and barnacles that settle on the shells of others and become encrusted are known as the epifauna. If a fossil shell is found with other organisms

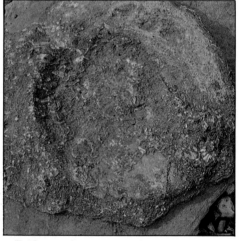

▲Epifauna of oysters on a fossil ammonite, 190 million years old, from the Lower Jurassic, Lyme Regis, England.

▲Fossil oysters cemented onto an ancient sea-bed or hard grounds. From Pskov, USSR, of Devonian age, about 400 million years old.

▲The fossil pond at Dura Den was excavated in the nineteenth century, but is 350 million years old. Among the fossils are lung-fish,

crossopterygian fish related to land vertebrates, and various placoderms or heavily armoured fish.

cemented onto it, then no matter how far removed it may be in relation to its original environment, the encrusting organisms are always considered life assemblages because they lived attached to the host shell, albeit in a different place – the association remains inviolate.

Often the remains of dead animals end up deposited on the sea-bed where there was perhaps a muddy sea-floor. Here they form a kind of artificial rock providing a miniature rocky island on which rock-dwelling creatures can find anchorage. Thus a muddy sea-bottom on which ammonite shells have come to rest can support animals that would be normally incapable of living in a muddy environment. Shells of dead ammonites, a death assemblage, provide a suitable environment for oysters and other organisms which cement themselves on the uppermost surface of the shells. From these associations it becomes possible to reconstruct the detailed conditions of the sea-floor and to recognize the differing micro-environments.

Hard grounds as the name implies are parts of the sea-bed where sediment has become cemented. It may be former rock or sediment that has been cut by wave action (a wave-cut platform) or even contemporaneous sediment that has become lithified. However they were formed, hard grounds make the ideal substrate for encrusting organisms such as oysters to settle upon.

These fossils are preserved not only in their life position, but the essential elements of the entire fauna are also preserved intact. As well as bivalves there are other encrusting organisms including serpulid worms and bryozoans. Also associated with hard grounds are organisms that are capable of boring into the substrate, such as rock-boring bivalves. At low water where the sea-bed is made up of a fairly soft type of rock such as chalk, limy mudstone, clay or shale, it will frequently be bored by piddocks which as they grow, rasp away the rock by means of the file-like ornamentation of the shell. Any sudden input of sands or mud over hard grounds is likely to kill the colonizing animals, since they are unable to move away (even wave action does not remove them, although the upper uncemented oyster valves are likely to be swept away). The creatures cemented or living in borings will be preserved as by far the most obvious of all life assemblages.

The chances of sand or mud-burrowing animals being preserved in their life positions are much less likely, but it does happen. The appearance of bivalves in a rock in their normal living position, the shells not jumbled up, but rather spaced out in the sediment is a clear example of a life assemblage. The contrast between life and death assemblages is seen at its most marked in the distribution of shells in the rock. In the life assemblage one normally observes cross-sections of the articulated shells, but unless one is careful they can be easily overlooked; whereas isolated shells that have been swept along into a heap, are very prominent and produce the sort of fossil death assemblage that is much more attractive to look at and seemingly more worthy of the collector's attention, but not by any means as informative.

▲ Petrified tree stumps in the Yellowstone National Park, Wyoming, USA, still in the position where they once grew.

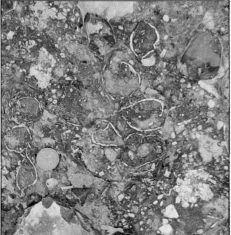

▲ A good example of a life assemblage: bivalves *Mya* or gaper clams Weybourne Crag, England, half a million years old.

TRACKS AND TRAILS

The tracks of ancient animals can be preserved as fossils. Although it is often impossible to identify the animals which made the tracks these fossils do illustrate the animals' behaviour.

Many sedimentary rocks which are quite devoid of fossils, may in fact be teeming with the evidence of past activity of living things. Such fossils are known as trace fossils; they are not the actual remains of animals, but a record of the animals' behaviour. It is these important traces that are most frequently overlooked by collectors, as one interesting example illustrates. On parts of the coast of Cornwall, south-west England, there occur considerable exposures of black Carboniferous slate, which was folded and refolded during the Armorican mountain-building period. The majority of these rocks show no signs of life and no fossil collector would look twice at them. However when some of these slates are split it is possible to see markings on the bedding surfaces. Generally these are noticeable only if the slab of rock is tilted so that the light catches the faint scratches. These scratches were formed immediately after the mud had been deposited and was either compacted or dried out sufficiently to enable delicate grooves to be marked on the surface. More importantly the grooves were actually preserved when buried by the next incursion of mud into the area. The study of scratches on a grey or black slate does not seem particularly promising, but a detailed analysis of the grooves suggests that they were made by limulids or king-crabs. The prints of what looks like a small four-toed foot originally lead to the mistaken belief that small four-footed vertebrates had walked over the muds. In fact king-crabs have a pair of appendages that act as pushers and it was these which scratched the trails on the rocks.

One king-crab trail which might not seem particularly significant at first glance comprises two sinuous curves which cross one another. The only way that this trail could have been produced with such regularity would have been for one king-crab to have been riding on the back of another. The one underneath was taking a winding path and the groove was made by its tail spine. On each

swerve the passenger crab would have swung further out and this gave the trackway of its tail a greater amplitude. Running along either side of the two sinuous curves are imprints where the backward-pointing spines of the head shield also grooved the sediment. These 300 million-year-old king-crab trails represent a mating pair moving on the muddy bottom. This is not the pattern today where king-crabs congregate at the water's edge and lay their eggs in the damp

sand. The fossil tracks give a hint that there has been a change in their breeding behaviour that would not otherwise have been suspected.

Another type of trace fossil, formed in a herringbone pattern, was discovered in South America in the 1840s. It was named *Cruziana* in honour of Andres de Santa Cruz, a former ruler of Peru and Bolivia. This fossil was not at first understood and similar tracks were discovered in Britain during the 1850s, which were

▲ A fossil showing the trail of a 150 million-year-old dying king-crab, from the Solenhofen lithographic limestone, S. Germany.

▲ *Cruziana*, the feeding furrow of a trilobite. From the Upper Cambrian, N. Wales, about 500 million years old.

▲Two sinuous out-of-phase trails in shale from the Carboniferous, Cornwall, England, 300 million years old.

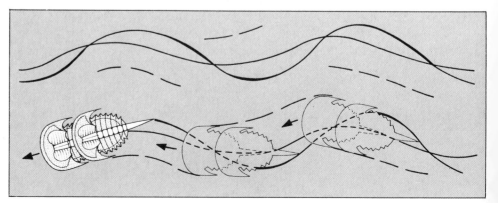

▲ A drawing of the fossil trails seen in the photograph above.

The reconstruction shows how the trail was formed during the mating of king-crabs.

interpreted as either sea-weeds or worm trails. These fossils are exclusively marine and commonly occur in fine sands and siltstones in rocks of between 570 and 500 million years' age. In many parts of the world the ancient sea floor was covered in these tracks. We now know that the trails were made by trilobites, primitive relatives of king-crabs, which, with their series of paired legs, ploughed the surface layers of the sea-floor sediment to extract organic debris for their food. The characteristic furrowing of *Cruziana* is associated with two other types of structure. One is rather like the print of a cloven hoof, *Rusophycus* and in Australia this was found with a fossil trilobite actually sitting in the trail, confirming the idea that such impressions were made when the trilobite rested. The *Rusophycus* impressions frequently pass into *Cruziana* furrows thus confirming that the latter were the foraging furrows. *Diplichnites* or 'double footprints' also occur in the same sediments and comprise a series of tiny parallel short grooves, the pairs being arranged at an angle to one another. These double prints are simply where the trilobites were no longer feeding on the sediment but moving about, perhaps seeking a more favourable organic-rich sediment.

These three seemingly different types of trace fossil simply represent different aspects of the trilobite's behaviour. The trilobites were among the most successful inhabitants of the sea-floor for 350 million years from the lower Cambrian to the late Permian. But trilobite trace fossils are rarely found after the Cambrian period because the primitive trilobites which characterized that period died out. The later trilobites developed many different ways of life but their sediment ploughing behaviour seems to have changed because they left no such tracks and trails.

In some cases it is possible to follow the evolution of trace fossils through periods of as long as 150 million years. The trace fossil *Dictyodora* shows how an unknown sediment-feeding animal became more efficient at obtaining its livelihood. In the early forms of the trace fossil (500 million years old) the animal lived a few millimetres beneath the surface and meandered through the sediment. In later forms it corkscrewed deep into the sediment and thereafter meandered. Later still the meandering path was much tighter and finally this creature simply corkscrewed in tight whorls deep into the sediment. Even if we can never identify the actual animal that produced these traces, it is nevertheless possible to document a gradual change in the behaviour of an inhabitant of the ancient sea-bed.

Trace fossils are usually found in what are popularly thought of as unfossiliferous rocks. A parallel situation can be seen today in such places as the Sahara desert. The overwhelming impression is of an environment quite incapable of supporting any animal life. Yet if one looks closely at the sand one is suddenly aware of the evidence of animal activity. Tracks show that there are beetles which tunnel just beneath the sand, emerge from their tunnels and walk across the surface; lizards crawl by in search of insect prey; birds hop across the sand hunting for insects; two lizards meet, scuffle and then part while a small rodent scuttles away. The seemingly lifeless desert is teeming with animal life.

Another inhospitable environment would appear to be the muddy regions of the sea-floor and the

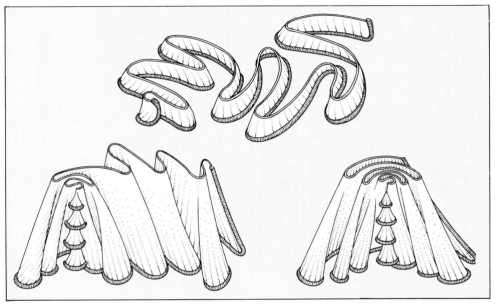

▲ *Dictyodora*, a trace fossil with a record which extends from the Cambrian to the Devonian, 500–350 million years ago.

▼ These are all fossil trails, made by trilobites.

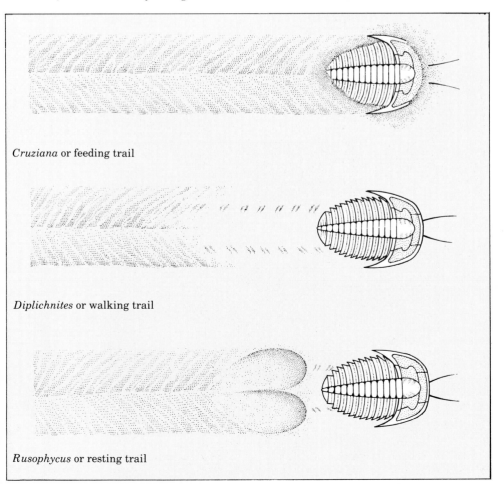

Cruziana or feeding trail

Diplichnites or walking trail

Rusophycus or resting trail

mudstones and shales which they become. On the southern coast of England, the area around Lyme Regis is justly famous for the spectacular fossils which litter the beaches, while the grey mudstones that dominate the shores hardly merit a passing glance from visitors. The ammonites and nautiloids, which capture the imagination of the fossil hunters are simply the shells of dead molluscs which have been swept into the area from elsewhere. The conditions of 190 million years ago on the sea-bed in this area are represented by the grey mudstones, and muds like the desert, seem rather inauspicious environments for living things. Certainly there is no sign of abundant fossils and the rock is a monotonous grey. But on closer examination it immediately becomes evident that the rocks contain traces of an environment swarming with many living things.

The grey calcareous mudstones or shales are not a uniform colour, but have lighter grey markings which represent the infilling by a different coloured mud from various types of burrow. By examining the markings exposed on the bedding surfaces and also in vertical sections, it is possible to reconstruct the actual three-dimensional patterns of the burrows and then work out whether the patterns represent dwelling or feeding burrows. The really difficult question is the possible identity of the animals responsible for the burrows. On extremely rare occasions an animal is found at the end of a burrow and it is assumed reasonably enough that such burrows were generally made by similar animals. Crustaceans, crabs and shrimp-like animals are noteworthy burrowers. The large calibre burrows are known to contain large shrimp-like crustaceans and by comparing the intricate tunnel systems of living crustaceans within the sediment, it is a reasonable inference to suggest that the sea-floor was inhabited by large numbers of such crustaceans. Tiny light grey spots belong to a system of sediment-feeding burrows and may well have been made by some kind of worm, which from a central point pushed out through the muds retreated somewhat, then branched off in a separate direction, only to retreat again, repeating the process. The net result is a pattern of radial burrows, many of which further subdivide. Since there is no sign of the offending worm it is only possible to work out how the burrow was formed

▲ Modern traces in the Sahara of a beetle burrowing and walking on the sand. Also shown are lizard's prints.

▶ A modern trail made by a limpet feeding off algal film on the rock surface with its rasping tongue.

▼A suite of meandering fossil burrows made by gastropods (large snails) as they wandered over the sea bed.

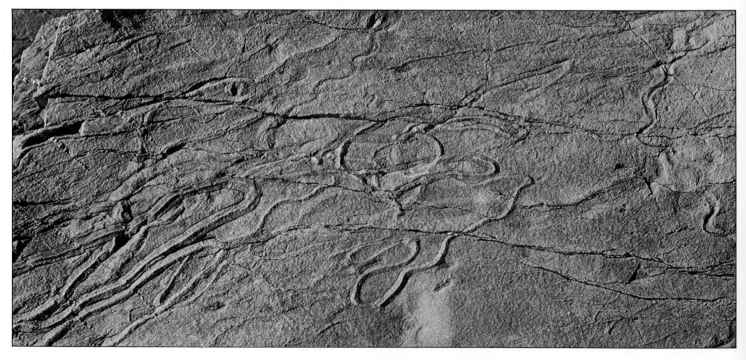

and hazard a guess at the identity of the culprit. Where the layers of sediment differ in colour it can frequently be seen that, at their interface, vertical pipes of one colour are incorporated into the underlying sediment, and in fact there may be a complete mixture of sediment so that only a few small patches of the original layering are discernible. This mixing is characteristic of rocks that are thought of as unfossiliferous and is entirely due to the organisms tunnelling in order to extract organic matter. Because it is caused by living things, this type of disturbance is known as bioturbation.

In many supposedly unfossiliferous, powdery sandstones it is possible to make out faint rings and parallel lines. When the sandstones are etched by natural weathering a pattern of tunnels and galleries can be observed, with walls made up of numerous pellets, which are slightly muddier than the surrounding sand. This trace fossil is called *Ophiomorpha* and was produced by a small shrimp-like crustacean, which sifted through the sediment pushing to one side the rejected particles through the mouth parts in the form of small pellets. There are a vast variety of similar crustaceans on tropical beaches today.

On northern beaches today tracks of living creatures, which may well become trace fossils, can be seen. The majority of the infauna (creatures living within the sediment) are different kinds of worms. Some inhabit simple vertical burrows lined with either slime or sand particles. This type of burrow implies that the animal's alimentary canal has a U-turn and waste matter is extruded at the entrance to the burrow. Perhaps more common are animals that live in U-shaped burrows, again different types of worm and also small crustaceans. In this burrow water is drawn in to the mouth at one end and waste material, such as the familiar worm-casts, extruded at the other. In conditions in which the water level rises and falls the actual burrows can be repositioned and in fact the entire system can migrate up and down; the trace fossil of one such burrow has been aptly named *Diplocraterion yoyo* after the children's toy. When sediment that has been occupied by such animals is itself buried by the influx of further sediment, the sands or muds retain the animal traces, providing clear evidence that the ancient sea-bed was a suitable environment for the support of animal life.

◀ Trace fossils from the Lower Jurassic, Lyme Regis, comprising wide burrows made by crustaceans and smaller ones made by worms.

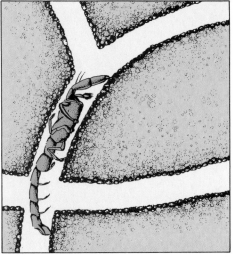

▲ The trace fossil *Ophiomorpha* is a pattern of horizontal and vertical burrows, made by the small crustacean shown in the diagram.

◀ These burrows are 215 million years old. They were found in Middle Triassic rocks in Tunisia.

CAVES

The fossils found in caves often provide unique evidence of the evolution of life in upland areas.

The fossil record has a built-in bias towards animals and plants living in or close to regions where sediments are being deposited; and the fossil record of land-living vertebrates is dominated by the larger types, partly because large bones are easier to find. But those animals which lived far from areas of deposition, the lowland regions, have an extremely poor fossil record.

Occasionally we are afforded the odd glimpse of life away from the main regions of sediment deposition. This is in limestone country – limestone is made of calcium carbonate. When rain falls it dissolves carbon dioxide from the atmosphere so that it reaches the ground as a weak solution of carbonic acid, which dissolves calcium carbonate. Joints and fractures in the rock provide initial routes for water to soak into. The effect of the acid is to dissolve the limestone along these same lines of weakness. Over tens of thousands of years, as the water accumulates, the rock beneath the water-level is dissolved by the acid, and large underground streams form within the strata. When the water-table falls, an empty space is left, above the surface of the water, which forms a cave. If the water-table falls still further, the floor of the cave itself may become a dry stream bed.

Once a cave has been formed and has become accessible as a consequence of the fall in water-level, it becomes available as a site in which animal remains can be preserved. When the water in a cave evaporates, the walls of the cave become covered by a layer of calcium carbonate, known as flowstone because it gives the impression of having flowed into position. When water drips from the ceiling, evaporation results in minute amounts of flowstone being deposited, forming an icicle-like pendant known as a stalactite. Where the drips land on the floor the reverse process takes place and slowly the calcium carbonate deposits build upwards to form stalagmites. Both these can cover and preserve animals that lived in the cave.

Animals enter caves for shelter. The extinct cave bears of Europe hibernated deep inside caves and during the long winter sleep the old arthritic bears died. Similarly many of the young, born late in the season, were not able to eat enough food and they too perished during the winter. Over the years their remains accumulated and in one cave system, in Austria, there are the remains of 30,000 cave bears. Caves were also inhabited by hyenas, and the remains of the food, which tended to accumulate in their dens, are often found.

While many of the remains found in caves are those of the mammals who used them for shelter, other fossils are found which have a different origin. In many instances animals have accidentally fallen down

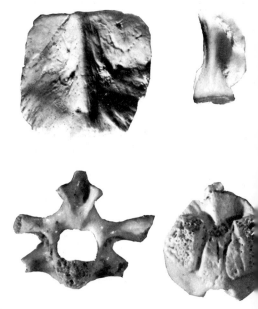

▲ Individual bones from the Middle Triassic caves, Gliny, Poland. Top, left to right: Archosaur bony armour and lizard quadrate (skull bone). Bottom, lizard cervical vertebra and base of brain case.

▲ Dark Devonian limestone (400 million years old) at Gliny, Poland showing burial of the cave system by the lighter-coloured Triassic sediment.

▶Section of cave and sediments
1 Cave floor covered by layers of sand and silt washed in by flowing water
2 Stalactites and stalagmites form in cave
3 Cave roof collapses
4 Remains of animals and sediment fallen into cave through roof fissure

5 Remains of animals washed into cave with sediment and of animals that lived and died in the cave.

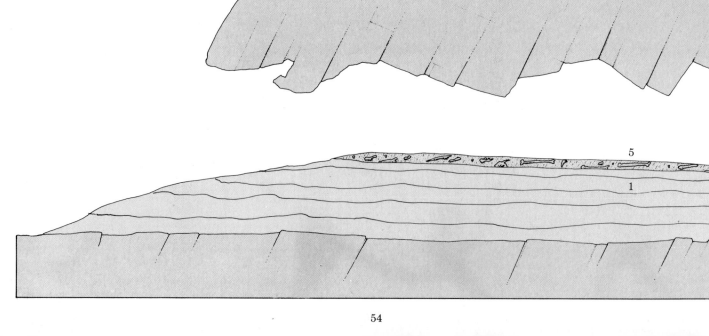

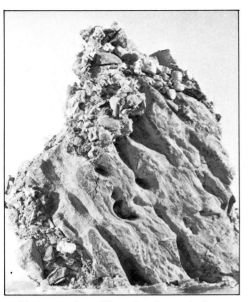

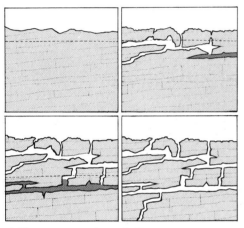

▲ Limestone from Gliny showing typical water erosion features and fragments of bones and pebbles that were deposited on the cave floor.

▲ How caves are formed. First water seeps into the rock through cracks; over time, acid in the water wears away the rock and when the water level drops, caves remain.

a sink-hole which leads into a cave system. Some caves are known where a few large animals have fallen down such holes to be followed by scavengers who have then been unable to get out. They have both subsequently been found as fossils. At the Joint Mitnor cave, Devon, England 4,307 bones have been recovered which make up 127 animals; the herbivores are mainly bison and deer, while the scavengers are chiefly wolves, hyenas and foxes.

A macabre assemblage of bones was found at Swartkrans in South Africa, where a vertical fissure of a cave contained innumerable hands, feet and skulls of baboons and early men – *Australopithecus* and the occasional *Homo erectus*. This curious selection of skeletal parts probably represents the remains of the meals of leopards which accidentally dropped into the cave.

Remains can also end up in caves when they are washed in by streams which flow into the cave system. Such remains are preserved in layered stream deposits formed on the floors of the caves

Limestone caves which form in upland regions may afford a rare glimpse of past life in such areas. Unfortunately, these regions always suffer erosion and there is no possibility of skeletal remains being preserved unless they can be buried rapidly. However there are a number of places where a fairly rapid transgression of the sea has first surrounded the upland areas, creating a series of isolated islands, and then covered them completely, thus preserving the entire system. Only

recently has this ancient landscape been exhumed and cave systems, filled with sediment and containing bones, insects and plant remains, discovered.

One of the earliest-known fossil cave deposits of this type comes from south-west Poland, from the Middle Triassic period (240 million years ago). As the sea from the ancient ocean of Tethys spread across most of Europe, it caused a line of hills running across south-west Poland, to become a string of islands. These cave deposits contained the remains of the immediate precursors of the dinosaurs, the first true lizards and small fish-eating reptiles related to the ancestors of the plesiosaurs. In Britain, cave deposits from the uppermost Triassic period have produced the very first true mammals known in the world: *Kuehneotherium* and *Morganucodon* from South Wales. Without such discoveries, we would never have known how far life had evolved on the land away from the lowlands and coastal regions. The variety of extinct reptiles recovered from a single cave deposit in the west of England has demonstrated that the variety of reptiles was about ten times greater than had previously been supposed. Although these faunas are difficult to relate to the mainstream of the fossil record, they are a timely reminder of the changes that were taking place in particular environments. In addition to these valuable discoveries the major study of cave fossils has been of those that formed during the last million years and contain remains of Ice Age mammals, including early man.

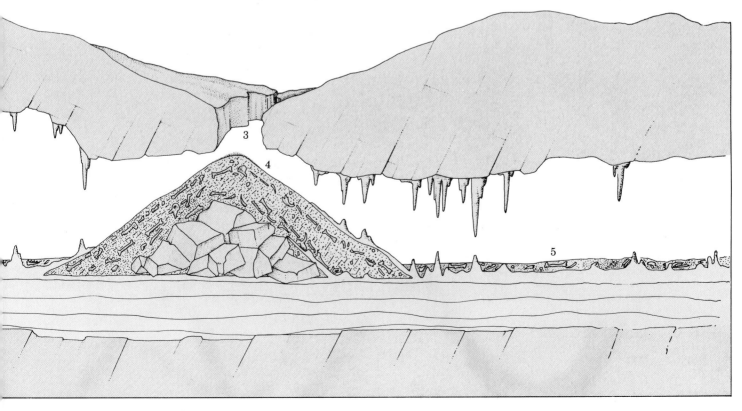

FUTURE FOSSILS

Studying a living community and tracing the fate of its remains
shows that the fossil record preserves a biased selection of past life.

As we have seen living things are likely to be preserved as fossils only if they end up in areas of sediment, usually along the edge of the sea. Animals which lived in or near these areas are most likely to be buried so it is not surprising that the fossil record is dominated by organisms which inhabited shallow coastal waters. The second essential attribute for fossilization is some mineralized hard part in the body of the organism.

The chances of preservation will also depend considerably on the part of the sea in which the animal lived. Those which live buried in sediment or which walk or crawl on the bottom have a high chance of becoming fossils, but the fate of the forms that swim or float at the surface is much more doubtful. Most are eaten by predators and scavengers, so that only fragments, scattered over the bottom, will be preserved. For the majority of animals and plants that inhabit the seas, the

likelihood of fossilization is remote. In any attempt to reconstruct a picture of life in the past, it must be remembered that a proportion of the community will not have left any record.

To investigate the different chances animals have of becoming future fossils, it is best to examine a rocky shore. Here rock pools occur between high and low-water-marks; when the tide goes down, pools of sea-water are left behind together with the animals and plants that inhabit the sea. Consequently, it is possible to study a small sample of the living marine fauna at leisure and in safety. All the major groups of living things can be identified with a guide to the seashore. Not all types of marine animals will be represented in one particular pool, but a reasonable cross-section will undoubtedly be present.

Among the most familiar animals to be seen in rock pools are sea-anemones, but not a single

anemone is known in the fossil record. Similarly, flatworms and sea-slugs or sea-hares, important components of the marine fauna, are never found as fossils. On the underside of rocks and in crevices there are tunicates or sea-squirts, which again are virtually unknown in the fossil record. There are, however, two exceptions; one, a small encrusting form that lived in colonies from the Cambrian period, 570–500 million years ago, and the other, a simple one from the Permian, 280–225 million years ago, preserved by virtue of the calcareous plates formed in its leathery tunic. Small carnivorous worms occur and these again are preserved only in very rare instances when they are overwhelmed by fine-grained muds.

Among other groups of animals to be found in rock pools, again often underneath rocks, are members of the echinoderms or spiny-skinned animals: the sea-urchins, brittlestars and

▲ A rock pool is the best place to study living marine organisms and assess their chances of becoming future fossils. In order to be preserved, an organism needs some mineralized hard part in or around its body. Soft-bodied organisms such as seaweeds, sea-anemones and sea-squirts leave no traces in the fossil record.

Future fossils

1 Cuttlefish bone
2 Egg case of dogfish
3 Razor shell
4 Mussel
5 Limpet
6 Whelk
7 Colony of barnacles
8 Winkle
9 Piddock boring
10 Glass bottle

starfish. These have skeletons of small calcareous plates which usually separate and become dispersed once the animal has died. However, if the sand where they lived is examined under a hand lens, it is usually possible to recognize the small plates and spines. These animals leave their remains to be preserved as fossils but they are so broken up and dispersed that they will generally be overlooked. Exactly the same sort of thing happens to fish. The individual bones and scales are separated and washed about until they become incorporated in the sediments. Again an examination of the sediment with a hand lens will provide evidence of the existence of fish. The other common inhabitants of rock pools are crabs and their relatives. The shells, or carpaces, are generally thin and delicate but some parts of their external skeletons, such as claws, are robust, and these are reasonable candidates to become future fossils.

It is also useful to observe what animals and plants occupy the surrounding rocks; a zone alternately covered by water and exposed to the air. Animals inhabiting these rocks must have some mechanism to prevent them drying out when they are exposed to the air and this is generally in the form of shells which can be tightly closed or held close to the rocks to which they are fixed. The most common animals, are barnacles which cement themselves directly onto rocks or other hard surfaces, such as empty shells, and remain there for life. Mussels, too, are common, fixed to the rocks by means of horny threads or 'beards'. Other common shellfish are limpets, winkles, whelks and top shells. Mussels, limpets and other sea-snails can no longer remain in their original position once they are dead, but their empty shells are swept away by the tides to accumulate with sediment, into which they become incorporated as future fossils. The barnacles are, in fact, much more likely to be destroyed by erosion rather than covered by sediments. If the rocks on which they are cemented become detached and rolled around by the tides the barnacles will be worn off and, when the boulders or pebbles become incorporated in sediment, there will be no sign whatsoever of their former encrustation. In spite of their tremendous abundance, barnacles leave only a sparse record in the form of fossils. The exceptions are those which settled on shells and were carried away with them. A thick shell is no guarantee of preservation as a fossil; an animal's habitat is also a critical factor.

It is also worth examining the materials that have accumulated on the strandline at the high-water-mark. This is the beginning of possible future fossiliferous deposits. There are likely to be shells of limpets, mussels and other sea-snails and possibly fish skeletons and crab carapaces, remains of birds and occasionally, mammals, such as seals. Shells belonging to animals which live buried in the sediment, such as cockles and razor shells, may have been washed up. Cuttlefish bones and the egg cases of skate and dogfish are swept onto the shore after floating about on the surface of the sea.

Seaweed will be plentiful on the strandline, and in and around the pools. It is infested with small, jumping sand hoppers and, if it has been there any length of time, will be rotting. Seaweeds disintegrate very quickly and are hardly ever found as fossils. Only the forms that deposit calcium carbonate on themselves will be preserved and these, in fact, have an excellent geological history.

Lost from fossil record

11 Sea-squirt
12 Sea-anemone
13 Sea-lettuce
14 Sea-slug
15 Seaweed
16 Plastic bottle

▲ A projected fossil record left by the rock pool and strandline. The fossil record has a built-in bias for creatures with hard shells or skeletons. To try to reconstruct a picture of what life was really like in the past it is necessary to put back the organisms we know existed, but which have not survived as fossils.

CHALK, FLINTS AND CORALS

Chalk is composed entirely of the fossil remains of microscopic marine plants. Flints and corals too can tell us much about ancient environments.

The carbonates in limestone and the silica in flint are extracted mainly from sea-water through the agency of living things, usually microscopic animals and plants floating on the surface waters of the oceans.

The most familiar of all limestones is chalk which covers vast areas of the present-day continents. Chalk was formed in the Cretaceous period (90 million years ago) at a time when the world's oceans extended to their maximum limits. At this period too, most of the land surface was low-lying and relatively minute amounts of sands and muds were being washed into the seas. Floating in the surface waters were unicellular microscopic algae, the coccolithophores. These algae form a hollow ball around themselves from secretions, comprising a series of circular plates which are themselves made up of smaller units. When the alga dies the plates separate and fall to the sea-bed, where they accumulate. Unfortunately these plates or coccoliths can be seen in detail only under very

high magnifications, such as given by the scanning electron microscope. In the Cretaceous period the warm, shallow seas accumulated vast volumes of coccoliths and these minute fossils make up the chalk which occurs so abundantly today.

The sea-floor of such a fine-grained deposit was inhabited by sponges, some of which had calcareous skeletons, while others had internal supports of minute spicules of silica. In both cases the mineral for their skeletons was extracted from the surrounding water. Swimming in the waters were fast-moving fish; in the chalk mud, worms lined their burrows with fish scales and heart-urchins also burrowed. Large bivalves rested on the sea-floor partly buried in the sediment. Whenever animals died there were few water currents likely to move and concentrate the remains, so that, although fossils are common in chalk, they are usually widely distributed rather than accumulated in shell banks; and in most cases they are not

far from their original habitat.

The other formations that are characteristic of chalk, and indeed much limestone, are flint and chert which are both made of silica (silicon dioxide). The difference between these two is that chert breaks with a platy fracture and flint with a conchoidal one, so-called because it has the

▲ A flint ring from the Cretaceous, 90 million years old, found in Cromer, England. It shows how flint formed.

▲ Chalk is composed of fossils, mostly coccoliths as shown in close-up in the second photograph. Together numerous coccoliths (microscopic calcareous plates) form the casing of a single alga.

curved, ribbed surface of a shell. The physical condition of flint is due to the silica being in a near glasslike form, ie not properly crystalline. Where vertical fractures or even bedding planes have continuous layers of silica, these are secondary, formed long after the chalk was deposited and are in fact often chert. Typical flints are in the form of irregular nodules. In some instances, where the chalk has been eroded, it is possible to observe the exposed bedding planes and the flints can be seen in the form of rings or paramoudra, sometimes with internal spokes of flint.

Some flints were formed within

▲ Growth rings on a Rugose coral from the Devonian, can help calculate monthly and yearly cycles of 350 million years ago.

the sediment as blobs of gel resulting from the solution of siliceous skeletons of sponges. Often flints form around shells or pieces of wood or any object that could serve as a nucleus. Indeed many form around sponges and wherever flints are found, say in modern river gravels or on beaches, there will always be some which have a concave surface, which is pitted or very irregular. This type often known as a false flint, was once hollow and contained a sponge; they are easily recognizable because they are lighter than the ordinary solid ones. Inside these flints is a fine white powder, known as flint flour. Under the microscope this is seen to be a mixture of sponge spicules, some shell-bearing unicellular creatures (foraminiferans), as well as two-shelled arthropods related to water fleas, the ostracods. The flint was formed around the living sponge and all the organisms that were associated with the sponge in life were similarly engulfed, so that when the organic matter finally disintegrated, the skeletal materials remained within the hollow flint.

In regions of the sea where the water is warm, shallow and clear, there are other types of limestone formation created by living things, but which, unlike chalk, are produced in

situ. These are coral reefs which can develop around shores or oceanic islands. As the sea-levels rise, the corals will grow to keep pace because they can live only at shallow depths, to produce atolls (circular coral islands). It was Charles Darwin who, during his voyage on the *Beagle*, first worked out this process.

One of the most interesting discoveries made in recent years by British palaeontologist Colin Scrutton and others, was that corals lay down a layer of their skeleton each day and it is possible to recognize not only daily deposits, but monthly and indeed yearly cycles of mineral deposition. Careful examination of the growth bands led to the surprising conclusion that the lunar month was longer than it is today; 400 million years ago it was over 30 days. Even more amazing was the realization that the number of days in the year about 570 million years ago was 428, and 400 million years ago was between 385 and 405 days. It had previously been suggested by astronomers that the friction of the tides is slowing down the Earth by about .0016 second per 100 years, and the evidence from fossil corals indicates that the days have lengthened appreciably during geological time, confirming their calculations.

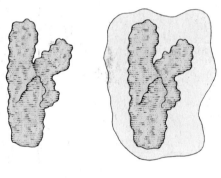

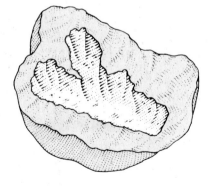

▲ The formation of false flint. Flint, initially a jelly-like substance, can form round a sponge. Years later when it is split apart, it contains the remains of the sponge inside.

◀ This sponge spicule is from inside a false flint. Spicules form the internal skeleton of a sponge.

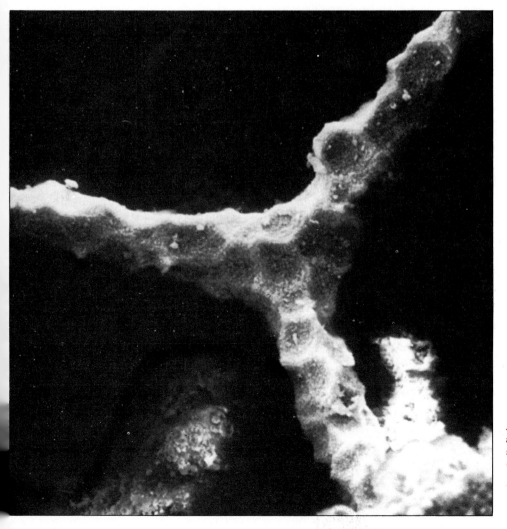

CHAPTER 3

HUNTING FOR FOSSILS

Man has been collecting fossils for the last
200,000 years. The most important finds are usually
the result of large-scale expeditions to remote corners
of the globe, but amateur collectors also continue to find
sensational material. In the eighteenth century
fossils were regarded primarily as curiosities of nature.
The eventual realization that fossils were evidence
of an antediluvian world led to organized expeditions
to document this past, most dramatically in the
discoveries of the denizens of the Age of Reptiles,
strange creatures of land, sea and air.

EXCAVATING GIANT SKELETONS

Great skill and care is needed to extract giant skeletons from the rocks and then transport them to a museum.

One problem common to all fossil-hunting expeditions and indeed to any palaeontologist discovering dinosaur remains is how to excavate giant skeletons. Today there is a standard technique used and the first task is to begin carefully uncovering the uppermost surface of the skeleton. This is done with a brush or a hammer and chisel depending upon the nature of the embedding sediments. As the bone is exposed it is coated with polyvinyl butyral soluble in acetone, or comparable solution, to protect it.

Once a specimen has been exposed, it is necessary to make a detailed and accurate map of the entire fossil. This information can be of vital importance when it comes to reassembling the skeleton, after it has been extracted, taken to the museum and prepared out of the rock.

Each bone has a number painted on it and lines are painted across fractures, so that if they come apart they can be accurately matched and stuck back together. The numbers are entered on the plan, and later painted on the crates in which the specimens are packed.

The original plan of the skeleton also shows the direction of north and the angle of the geological strata, as these can be especially important in determining the history of the specimen. The posture of the skeleton can be indicative of the animal's death and the subsequent fate of the skeleton. A single skeleton with the neck drawn back and the tail raised suggests that the animal died in an arid environment, because the flesh rotting away, the drying of the tendons and ligaments shrinking caused the head and tail to draw up. Where the bones or skeletons are mixed up and the long bones point in more or less the same direction, suggests that they were swept downstream during a flash flood. Where the skeleton is simply disarticulated and the bones mixed up this may be the work of scavengers. A detailed and accurate map also enables the palaeontologist to determine whether there are the remains of more than one kind of animal.

The map of the carnivorous dinosaur, *Tarbosaurus* (found on the 1964 Polish-Mongolian expedition) shows a rigid back from the shoulder extending partway down the tail and the flexible neck, and suggests that the mounted skeletons of similar dinosaurs in New York, Moscow and Ulan Bator are probably inaccurate.

Once the fossil has been mapped in its original position, it is removed from the rock, packed and transported to the museum. The bones are brittle, so all this has to be done with great care. One of the less happy realizations that came about during the great

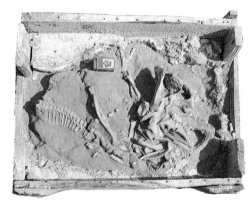

▲ Partial skeleton of the small herbivorous dinosaur *Protoceratops*, encased in plaster and crated.

▲ Diagram showing the exact position of the skeleton of *Tarbosaurus* in the sediment, providing crucial evidence on the posture of bipedal carnivorous dinosaurs.

▼Stages in extracting and packing a fossil

Carefully uncover fossil using mounted needles, or hammer and chisel depending on its size.

Dip cloth bandages in plaster of Paris and lay lengthways, then crossways, on the fossil.

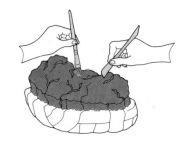

The encased fossil is broken off the pedestal and turned over and replastered.

Paint, then cover with wet newspaper, filling in all the cracks.

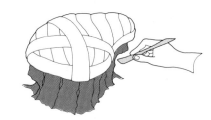

When the fossil is encased in plaster, the underlying rock is cut away, leaving a rocky pedestal.

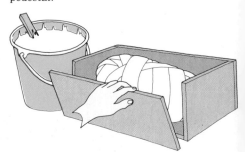

The specimen is crated and packed with more plaster, ready to be transported from the site to the laboratory.

American dinosaur bone rush of the 1870s was that no matter how carefully the fossils were packed, by the time they had been bumped over rough ground in horse-drawn vehicles, the hard-won trophies had been reduced to useless rubble. The solution to this problem is attributed to the palaeontologist, E. D. Cope, who disliked eating rice, the staple diet for expeditions to the American West. He had the rice boiled up into a paste in which he soaked the strips of sacking which he used to wrap up the bones. The paste set hard and protected the bones from breaking up during the journey back to the museum. This technique proved so effective that it was adopted by all subsequent expeditions. It was soon realized that plaster of Paris was more efficient than rice, and canvas bandages were used rather than sacking. But fundamentally this technique has hardly altered since the days of Cope.

Where large bones are to be removed from the ground, the first step is to expose them so that most of the specimen is uncovered. With a limb bone, more than half the diameter is exposed. It causes untold damage if plaster is applied direct to a fossil – there has to be a layer separating the bone from the plaster. Wet paper, such as newspaper or toilet paper, is applied first of all to the specimen and packed into any cracks or crevices and also onto the undercut parts of the bone. Strips of canvas, burlap, or plasterer's scrim are dipped in plaster and applied lengthways along the exposed bone and smoothed into position by hand. When this is done, the process is repeated with the strips running across the bone. Sometimes it may be necessary to apply wooden splints to the bone both for added support and ease of subsequent handling. Once this stage has been reached, the underlying rock has to be partly cut away so that the bone rests on a rocky pedestal. It may help to drill holes in the wooden support to thread dry bandages through, tying them over the top of the specimen. Then the specimen can be cut off the pedestal and rolled over. The dry bandages and the rock can be carefully removed to expose the rest of the bone. Again wet paper is applied and plaster bandages laid along the length and then transversely until the bone is completely encased.

On many major expeditions a carpenter's workshop is set up, so that wooden crates can be built around specimens. Only one side of the specimen need be exposed and, after the initial protective covering, plaster is poured over it and the box turned over so that the bone is taken back encased in the underlying sediment. This technique is hardly suitable for large dinosaur skeletons.

At the end of the 1960s, a modification of the standard plastering technique was invented making use of modern plastics. Instead of wet paper, the exposed specimen is covered in aluminium cooking-foil and then covered in polyurethane foam. Some collectors have combined both techniques so that foam is used for the initial covering, but once the specimen has been turned, the normal plaster bandaging is used. The main problem with expanded plastic foams is that the solutions may be poisonous, and the fumes given off can be toxic in an enclosed space, especially if anyone is smoking. With plaster bandaging there are no such hazards. The newer techniques should be left to the most experienced of museum technologists.

▲ A wooden frame is built round the plastered fossil, filled with plaster, undercut and turned over with the aid of a truck.

◄ Camp of the Polish-Mongolian expedition in Altan Ula, Gobi Desert. Erosion and rock exposure make it ideal fossil-hunting terrain.

◄ The plastered fossil may have to be loaded directly onto a truck to be taken back to camp for crating.

▼ The crated specimen has to be manhandled onto the truck for transport from the field to the laboratory.

FOSSIL HUNTING IN THE GOBI DESERT

The rocks of the Gobi Desert have yielded an unsurpassed collection of fossil remains to three major series of expeditions.

Palaeontological interest in Asia first started at the end of the nineteenth century when there was speculation that the origin of modern mammals and even mankind was to be found there. The attraction of the Gobi Desert in particular was that it was evidently a region that had not been inundated by the sea so that land animals were likely to be preserved there. This possibility fascinated Roy Chapman Andrews of the American Museum of Natural History and in 1922 he led the first expedition to hunt for fossils in the Gobi Desert.

The American expedition arrived in the spring of 1922, hoping to find the traces of mammals in the Cretaceous rocks. The expedition comprised a 26-man team which travelled in cars, while a camel train carried the expedition's supplies and fuel for the cars. Only a few mammals were found but the scientists were rewarded by the first dinosaur remains known in Asia north of the Himalayas. These were of disarticulated duckbill

dinosaurs as well as giant flesheaters. The perfect skull was found of a very primitive horned dinosaur, which was the long-sought ancestor of *Triceratops*, and was named *Protoceratops andrewsi* in honour of the expedition's leader. But perhaps the most exciting discovery was of some dinosaur eggs proving beyond all doubt that the dinosaurs were egg-laying.

When the autumn weather set in the Americans were forced to withdraw, but they returned as soon as possible in the following year. Over two years, 1923 and 1925, they excavated a unique collection of *Protoceratops* remains consisting of 75 skulls and 12 complete skeletons showing all growth stages from hatchling to 2 m long adults. More fossil eggs were found in large numbers, belonging to *Protoceratops* and other reptiles. Some of these eggs were arranged in concentric circles in 'clutches' preserved in the position they had been laid. Other new dinosaurs discovered were the very rare

Psittacosaurus and *Oviraptor* (egg-stealer), so-called because the skeleton was preserved on top of a cluster of *Protoceratops* eggs. These amazing dinosaur finds attracted much publicity, but perhaps the most important discoveries made were the first placental or modern mammals such as *Deltatheridium* and *Zalambdalestes*.

The Americans visited Mongolia again but were unable to excavate and transport some of the giant-sized material. It was not until 1946 that the Gobi Desert was again visited by palaeontologists. In that year a reconnaissance trip was made by the Academy of Sciences of the Soviet Union led by Professor I. A. Efremov. During their reconnaissance the Russians crossed the Gobi Desert twice and discovered a cemetery of large dinosaurs. They then mounted two full-scale expeditions in 1948 and 1949. The overall weight of the collections of dinosaur remains made in those three years amounted to 120 tons. Several

▶ The skull of a bone-headed dinosaur weathering out from the rock in the Gobi Desert.

▲ These small skeletons, known as the 'fighting dinosaurs', show the plant-eater *Protoceratops* in the embrace of the carnivorous *Velociraptor*. They were found at Tugreeg in the Gobi Desert during the 1971 expedition. The drawing is a reconstruction of what probably happened 70 million years ago.

complete skeletons of the giant bipedal carnivore *Tarbosaurus* (very similar to *Tyrannosaurus*) and the duckbilled *Saurolophus* were excavated and are in the Palaeontological Museum of the Academy of Sciences, Moscow.

The third and most productive series of expeditions to the Gobi Desert, 1964–1971, were organized jointly by the Poles and the Mongolians. Five Polish scientists arranged a reconnaissance trip in 1963 to locate the main bone-bearing sites and any geological formations which might yield remains. They were then able to work out the equipment required to support a full scientific team as well as the excavation itself together with the packing and transport of material. The final list was over 20 pages and included a six-wheel drive lorry, capable of negotiating sand dunes, and specially designed tents to withstand the harsh conditions of the desert. The leader of the joint expeditions was Professor Zofia Kielan-Jaworowska.

These expeditions excavated important dinosaur skeletons and other fossils. One of the first skeletons to be discovered in 1964 was another *Tarbosaurus*. A sensational find in 1971 were two complete skeletons – a small herbivorous *Protoceratops* locked in a death embrace with the carnivorous *Velociraptor*. These specimens are the only direct evidence of dinosaurs fighting. A beautifully preserved near complete skeleton and spiky armour of an ankylosaur was named *Saichania* based on *saichan*, the Mongolian word for beautiful.

As far as the palaeontologists were concerned, probably the most exciting results were many mammals contemporary with the dinosaurs such as *Kamptobaatar* (a rodent-like animal) and *Kennalestes*. These finds were very important because they gave a wider perspective on mammalian evolution at a very early stage.

The results of any major palaeontological expeditions always arouse tremendous public interest. By far the biggest attraction are the mounted skeletons of dinosaurs, which involve considerable engineering skill either in making unobtrusive metal frames to support the bones or in casting. One of the first complete dinosaur skeletons to have several casts made of it was *Diplodocus carnegiei*, named after the American steel magnate Andrew Carnegie who had financed the expedition to Utah where it was found. Carnegie was so pleased with this splendid specimen bearing his name that he arranged for plaster casts to be distributed to museums throughout the world. Today casts are made of fibreglass, because it is so light, while the real bones are kept by museums for detailed study.

The excitement of seeing their collected material mounted and on display to the public is certainly gratifying to the members of an expedition. To the scientific community, however, the public display is not paramount – what matters above all is that the scientific results of their work are communicated throughout the world.

Scientific results are published in scientific journals or monographs. In the case of the Polish-Mongolian expeditions nine monographs have been published to date. When an expedition uncovers evidence of animals new to science, there is much detailed research to be done. The new material must be prepared, described and then assessed in relation to similar types of material discovered in other parts of the world. This may take several years; indeed, the monographs on material collected by the German expeditions to East Africa before the 1914–18 war were still being published in the 1950s. In this type of journal many aspects of an expedition's work are recorded including detailed logging of the rock sequences.

The scientific results of the Polish-Mongolian expeditions to the Gobi Desert are contained in 9 monographs and more are in preparation. They describe in detail the fossils, most of which are new, and the rock and sedimentary structures.

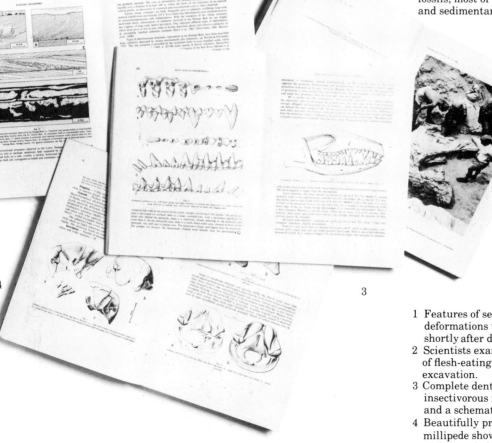

1 Features of sediments showing deformations that took place during or shortly after deposition.
2 Scientists examining the complete skeleton of flesh-eating dinosaur *Tarbosaurus* during excavation.
3 Complete dentition of a small placental insectivorous mammal *Kennalestes gobiensis* and a schematic restoration of its skull.
4 Beautifully preserved fragments of a millipede showing details of isolated portions of the exoskeleton of the head.

AMATEUR FOSSIL HUNTERS

Amateur fossil hunters have made many major finds. Chance discoveries have also contributed important finds to the fossil record.

Major fossil finds are sometimes made by school-children who notice something different in the rocks, because, unlike the experts, they have no preconceived notions about what should or should not be there. The amateur fossil hunters, who have traditionally provided the materials on which the professionals have built their scientific reputations, have two basic attributes: an eye that can pick out the faintest fossil trace, and the patience and persistence to spend days, weeks, months, even years, carefully extracting specimens from the rocks.

Since the eighteenth century there has been a tradition of fossil collecting in Britain, when the landed gentry filled their cabinets with curiosities, and country people were continually on the lookout for objects which they could sell to visitors. The area around Lyme Regis in Dorset has highly fossiliferous rocks and there was a carpenter named Richard Anning living in the town, who supplemented his income by selling fossils. In 1810 he died and his widow and children decided to continue the fossil-selling business. That year the eldest son Joseph saw what seemed to be the head of a strange animal protruding from the cliff; but it was his 11-year-old sister Mary who took a leading part in organizing the gradual removal of the remains. It was the first skeleton of the sea creature *Ichthyosaurus* to have been discovered and was sold for £23. When she was older Mary discovered the first *Plesiosaurus*, another sea reptile, which was sold to the Duke of Buckingham for £100, and later one of the first flying reptiles, the pterosaur *Dimorphodon*. Mary Anning (1799–1847), the 'fossil woman of Lyme Regis', became famous for her discoveries.

Because the public's imagination is caught by the interest of large-scale expeditions to remote regions, it is always assumed that the original localities of historic discoveries have been denuded of all accessible remains. This is not the case. The erosive power of the sea, especially during the winter storms, continually exposes new sections with fossils and would go on to destroy them in time, were it not for

the existence of small bands of dedicated enthusiasts. Today at Lyme Regis Robert Langham, his son Peter and David Costain carefully monitor the coastline, retrieving every bone-bearing piece of rock that becomes exposed before the natural processes of erosion destroy it for ever. It is always a race between the fossil hunters and nature. In wet wintry conditions, huge boulders, or sometimes thin broken layers of limestone and shale, are brought back to the workshop, a room off the kitchen where they are prepared. Here, for instance, a thin bed of limestone composed of delicate fronds of sea lilies would be carefully pieced together like a jigsaw.

Sometimes a calcareous boulder is found which contains a reptile skeleton, of which the only visible sign is a small fragment of bone seen in cross-section. Enormous care, patience and some expense are needed to successfully uncover the skeleton. The boulder is partially supported by a

wrong method

correct method

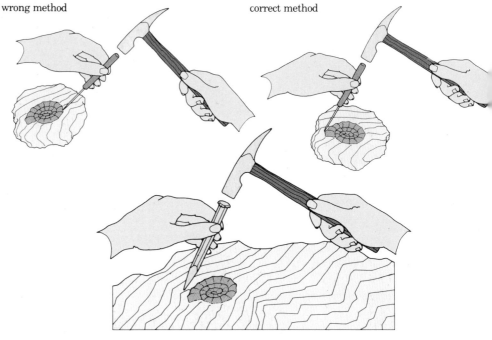

▲ How to chisel a fossil out of rock
It is important to hold the chisel vertically so that vibrations set up by the hammering affect any lines of weakness, causing the rock to split neatly revealing the fossil. Do not point the chisel towards the fossil, as it can easily be damaged. For finer preparation, use a mounted needle, held in the same way.

▲ The living-room of amateur palaeontologist Peter Langham is dominated by his extensive fossil collection.

plastic support or is simply coated by amyl acetate and the entire specimen immersed in a weak 20% solution of acetic acid. This acid attacks calcium carbonate but not the calcium phosphate mineral of bones and teeth. After several days, as the limestone dissolves, pieces of bone become exposed. The boulder is taken out of the acid and placed in water for a few days, then dried and the exposed bone coated with acetate or another coating. When this hardens the boulder is replaced in the acid until further bone is revealed. The process of washing and coating is repeated again and again until the specimen is fully exposed, or until it reaches a stage where, if the process were continued, the skeleton would disintegrate or collapse.

The chemical treatment of a complete skeleton can take many months of painstaking work. The standard of preparation of the Langhams and Costain is surpassed only by the combined professional skills and resources of the palaeontological laboratories of national museums. To the surprise of the scientific community, this group of amateur fossil hunters continues to provide new discoveries together with new details of the anatomy of familiar reptiles, such as a baby ichthyosaur on loan to the University of Reading which has preserved parts of its skin as well as its stomach contents. Lyme Regis, famous for its discoveries of prehistoric reptiles, now has the results of the Langham-Costain collections on public display in the Lyme Museum; one remarkable specimen even has details of the alimentary canal preserved. The homes of the Lyme Regis group are veritable museums, their walls and shelves covered with superb specimens. In 1981 a completely new long-necked plesiosaur found by them was formally described under the name *Kimmerosaurus langhami*, and in the same year the discovery of the third specimen in the world of the primitive armoured dinosaur *Scelidosaurus* was announced.

The role of the amateur ranges from the 'rock-hounds' that descend locust-like on fossiliferous localities through to the individual enthusiasts making original scientific contributions. Some of the major figures in palaeontological research of recent years turn out to be amateurs who have devoted their spare time and energy to specializing in areas neglected by the professionals. For example, F. C. Stinton of England has become the world authority on fish otoliths – small calcareous bodies from the inner ear. In France, one of the international experts on trilobites is the medical doctor Jean Pillet of Angers. In Germany, Professor Dr Wilhelm Stürmer, of Siemens Medical Division, Erlangen, has achieved world reknown for his radiographic studies of fossils from the Lower Devonian roofing slates of Hunrück in the Rhineland (see p 42).

The advent of mechanical quarrying has resulted in the unintentional destruction of much fossil material. However, the quarries of the London Brick Company, England continue to provide important skeletons due entirely to the co-operation and enthusiasm of the management and work force. A large-scale restoration of the plesiosaur *Liopleurodon* adorns the entrance to the main offices.

The scientific community today depends on the dedication of enthusiastic amateurs and on the acute observation of quarry and clay-pit workers to ensure that precious fossil material is not destroyed.

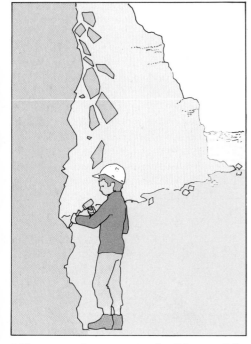

▲ Never try to hammer out a fossil from a cliff face as this could bring down tons of rock.

▲ Mr W. W. Wolfe in the 1950s uncovered part of the hip-girdle of a giant plesiosaur buried in a clay-pit in Stretham, Cambridgeshire, hence its name *Stretosaurus*.

▲ Dave Costain carefully pieces together a thin limestone band comprising numerous perfectly fossilized crinoids (sea-lilies).

THE FIRST FOSSIL COLLECTORS

**Fossils were collected by cavemen, but the Greeks were the first
to realize that fossils were remains of living organisms. Subsequently,
many people preferred to believe they were formed in the earth.**

Ammonites, with their coiled shells, and sea-urchins, with their five-fold markings, are among the more striking fossils which are commonly collected. They appeal both to people who have a knowledge of palaeontology and to those who simply enjoy the beauty of natural objects.

Some 200,000 years ago early man, *Homo erectus*, collected fossils but he could have had little notion of their meaning. At this time the technical equipment of man was an all-purpose tool, the biface hand axe of the Acheulian culture (see pp 170–1). A hand axe has been found in Norfolk, England, which was fashioned from a block of flint with an impression of the bivalve shell *Spondylus*. This had been worked in such a way that the fossil shell ended up exactly in the centre of the tool. The deliberate positioning of this fossil would indicate the dawning of mankind's aesthetic sense. Another

hand axe from Swanscombe, Kent, England, shows a comparable centrally placed fossil, in this instance the sea-urchin *Conulus* with its five-fold symmetry. There are even chert tools made from fossil corals preserved in chert found in gravels from Swanscombe. They must have been carried there since the nearest outcrop of coral-bearing chert is 200 km away. This evidence predates cave carvings and paintings by some 180,000 years.

The advent of Cro-Magnon man *Homo sapiens*, some 20–30,000 years ago, is notable for the rich paintings, carvings and sculptures (see pp 172–3). Men and women also decorated their bodies, clothing and hair with ornaments of various kinds and among the materials collected for such purposes were fossils – gastropods with high-spiralling shells and trilobites which were pierced and made into necklaces and amulets. Bear and wolf

teeth, as well as beads of mammoth ivory, were also made into necklaces. With shells it seems that Cro-Magnon man did not distinguish between the fossil and the modern. However, fossils were collected and carefully conserved by early man, not merely to embellish his hand axes but also his person.

It is not until the time of the ancient Greeks that we begin to have written evidence of how man interpreted fossils. Xenophanes of the sixth century BC was among the first to note the existence of fishes and marine shells preserved in rocks of inland mountains. At much the same time Pythagoras and others sensibly concluded that these fossils showed that the mountains were once covered by the sea. Herodotus (484–420 BC) wrote a book on Egypt in which he recorded the presence of marine shells far inland and concluded that they had been left there by the sea. These

▲ An Acheulian flint hand axe decorated with one valve of the Upper Cretaceous bivalve *Spondylus spinosus*, from West Tofts, Norfolk.

▲ An Acheulian flint hand axe decorated with the sea-urchin *Conulus* from Swanscombe, Kent.

▶ Four Cro-Magnon necklaces from Czechoslovakia. The cylindrical beads are carved from the tusk of a mammoth, the shell necklace is made from gastropod shells and the other two strings are made from fox, wolf and bear teeth.

interpretations did not become the accepted view. Indeed it took about 2,000 years for these early ideas to be re-established in Western thought.

Theophrastus of Lesbos (370–288 BC) was of the firm belief that fossil bones were formed from some 'plastic virtue' or force latent in the earth. This view was developed from Aristotle's theory of the spontaneous generation of living organisms. Theophrastus's interpretation was the one that truly appealed to the thinkers of the period and was firmly held for many generations. Pliny the Elder (AD 23–79) in his *Natural History* also promoted the notion, 'that there be bones growing in the earth'. For fossil sharks' teeth he had a more intriguing interpretation: 'Glossopetra resembleth a man's tongue and groweth not upon the ground, but in the eclipse of the moon falleth from heaven.' These ideas were more in tune with the magical and mystical interpretations of phenomena prevalent at the time and the commonsense approach of Herodotus did not appeal.

In other civilizations fossils were observed and their details carefully recorded. During the fourth century the Chinese noticed fossil formations which they called 'stone swallows'. We now know that these were spiriferid brachiopods (Palaeozoic shellfish), which have the appearance of swallows' wings. (In south-west England similar fossils are known as the 'Delabole Butterflies' for the same reason.) In AD 1133 a careful description of fossil fishes (now known to be *Lycoptera*) and their occurrence in the rocks, including their form of preservation, was published in China. The author noted that these fossil fishes, with their perfectly preserved scales, were collected, and their popularity was such that, 'the local people make falsifications of these fish by painting stone with lacquer . . .' There must have been a considerable demand for these fossils if forgeries were made.

The Chinese descriptions indicate that it was realized that fossils were the genuine remains of past life. In this regard their views contrasted with those generally held in Europe as late as the eighteenth century. The existence of rock in the form of shells – natural casts of shells (see pp 38–9) – seemed to support the belief that they were 'formed stones' produced by plastic virtues latent in the earth or *vis plastica*. This interpretation survived until the end of the seventeenth century in Britain. Robert Plot (1640–96, the first Curator of the Ashmolean Museum, Oxford) in his *Natural History of Oxfordshire* (1677) described and illustrated numerous fossils and discussed their mode of origin, concluding that they were formed stones. Yet in some instances his conjectures were correct, as when he described the distal end of a femur preserved in a rock and concluded it was a real bone now petrified. However, he thought it was probably from a giant man. In fact it was the first dinosaur bone (from *Megalosaurus*) ever to have been described. Subsequently R. Brookes, in his book *The Natural History of Waters, Earths, Stones, Fossils and Minerals* (1763), copied Plot's description in an illustration which he labelled *Scrotum humanum*. The French philosopher J. B. Robinet (1735–1820) described the same fossil as a 'stony scrotum' in his book published in 1768. Robinet's interpretation was that such objects were nature's attempts to form human organs in an effort to create a perfect human type.

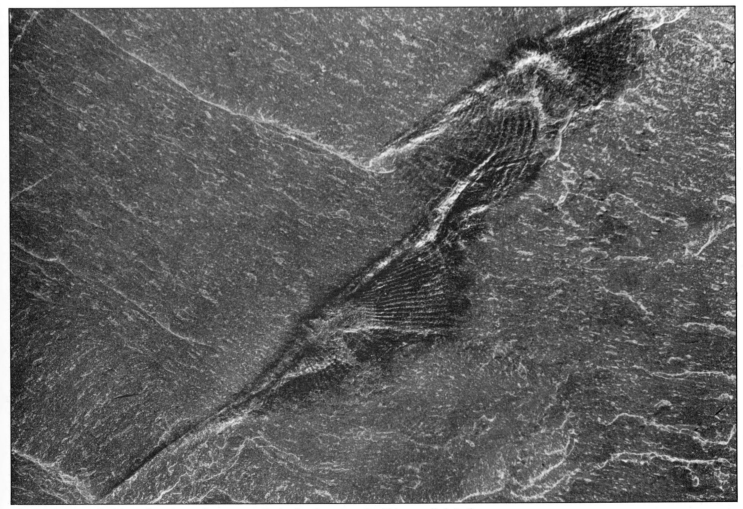

▲ 'Delabole Butterflies' from slates in south-west England. These fossils are actually spiriferid brachiopods (shellfish) which have been distorted.

In the fourth century the Chinese called similar fossils 'stone swallows'.

THE SIGNIFICANCE OF FOSSILS

Another erroneous theory about fossils was that they had been deposited by the Flood. Although Leonardo da Vinci had argued forcefully against this, the idea persisted until the eighteenth century.

The explanation of fossils in terms of *vis plastica*, or plastic virtues, lasted for two thousand years. Such interpretations were mercilessly ridiculed, even while they were current, by Robert Hooke (1635–1703) in 'Discourse on Earthquakes', later published in *The Posthumous Works of Robert Hooke* (1705). He understood that ammonites were the remains of marine shellfish and correctly compared them with their living relative, *Nautilus*. By the time John Morton's *Natural History of Northamptonshire* was published (1712), fossils were simply interpreted as the remains of once living organisms and allusions were made to the comparisons with their living representatives. Morton's book was modelled on Plot's basic plan, in which the illustrative plates were paid for by individual patrons; this fostered the tradition of each plate being dedicated to an important benefactor.

The view that fossils were the genuine remains of living creatures had been put forward by the Greeks in the sixth century BC. Later, this became more easily accepted when the layers of sediment and the fossils they contained were attributed to the action of the Universal Deluge, the Biblical Flood. This explanation of the presence of seashells in rocks of high mountains was first advocated by the early Christian writer Tertullian (AD 155–222): 'Yes, and the whole earth was changed once, being covered by all the waters. To this day, sea conchs and tritons' shells are found as strangers on the mountains, desiring to prove to Plato that the heights have once flowed with water.' This diluvial explanation attracted its most enthusiastic adherents during the late seventeenth and early eighteenth centuries, notably the English physics professor, John Woodward (1665–1728), who was ironically nicknamed 'The Grand Protector of the Universal Deluge'.

A further enthusiastic diluvialist was the Swiss Johann Scheuchzer (1672–1733). He is best remembered for describing a fossil giant salamander

as 'the bony skeleton of one of those infamous men whose sins brought upon the world the dire misfortune of the Deluge.' He named this 'sinner' *Homo diluvii testis*, or man, witness of the Flood.

The discovery of fossilized ripe fruits from the London Clay of the Isle of Sheppey was described by J. Parsons in 1757 and taken as evidence that the Flood took place in the autumn. Today the 'scientific' Creationists continue to interpret the geological sequence and

all the fossils as having been formed during the Flood. They think that the record of the rocks demonstrates evidence of catastrophic events and that fossils similarly provide incontrovertible proof of the sudden violent demise of living things,

The preservation of large numbers of skeletons swept together in great heaps certainly can give an impression of a catastrophic end. There are undoubtedly many instances in the geological record of sudden incursions

▲ The Swiss Johann Scheuchzer (1672–1733) was an avid fossil collector, and this is the frontispiece to the catalogue of his museum, 1716.

or floods across entire regions, known as marine transgressions. Flash floods are all too common in arid and semi-arid regions. But in drawing conclusions one has to take a balanced view of these instances, together with the numerous examples of slow gradual changes in the rocks and organisms dying in their habitats.

In any case this particular question was dealt with by Leonardo da Vinci (1452–1519) and his clear interpretation can serve to settle the issue: 'If you say that the shells which are found in our times within the confines of Italy, far away from the sea, at such great heights, were left there by the Deluge, I answer that if you believe that this Deluge rose above the highest mountain by seven cubits, as he who measured it has written, these shells, which always inhabit the sea-shores, ought to be found lying on the mountain sides and not so little above their bases, and all at the same horizon, layer upon layer.

'. . . shells have been carried empty and dead by the waves, I say that the dead could not have been far removed from the living, for in those mountains are found those which had been living, recognizable by having their valves united, and all are in a layer where there are no dead ones; and a little higher are found, as if heaped up by the waves, all the dead ones with their valves separated.

'. . . In the mountains of Parma and Piacenza multitudes of rotten shells and corals are to be seen, still attached to the rocks . . . And if you were to say that such shells were created, and continued to be created in similar places by the nature of the site and of the heavens, which had some influence there – such an opinion is impossible for a brain capable of thinking, because the years of their growth can be counted on their shells, and both smaller and larger shells may be seen, which could not have grown without food, and could not have fed without motion, but there they could not move.

'And if you wish to say that it was the Deluge which carried these shells hundreds of miles from the sea, that cannot have happened, since the Deluge was caused by rain, and rain naturally urges rivers on towards the sea, together with everything carried by them, and does not bear dead objects from the sea shores towards the mountains. And if you would say that the waters of the Deluge afterwards rose above the mountains, the movement of the sea against the course of the rivers must have been so slow that it could not have floated up anything heavier than itself.

'And even if it had supported them, it would have scattered them in various places when it subsided. But how can we account for the decayed corals, which may be found any day in the neighbourhood of Monte Ferrato in Lombardy, attached to the boulders uncovered by the river currents? These rocks are all covered with colonies of oysters, which as we know, do not move about, but are always attached by one valve to the rock.' (Leicester MS.)

The extensive range of Leonardo's interests is legendary, and fossils did not escape his scrutiny. Notwithstanding the strong prejudices of his day, Leonardo relied on direct observation and reasoned analysis to reach an understanding of the organic origins of fossils and how they came to be in the rocks. His accuracy is astounding. He courageously refuted the theory of the Flood and in so doing anticipated the modern approach to interpreting the history of the Earth.

▲ Engravings from *The Natural History of Northamptonshire* (1712) by John Morton. These show many fossils including ammonites, sea-urchins and vertebrae.

LOST SPECIES

In the eighteenth century, bones were found in Siberia and North America which proved to be of unknown animals.

The eminence and reputation of Leonardo da Vinci enabled him to hold what were considered to be seriously heretical views, although the Bishop of Florence felt constrained to point out to him that, 'the Almighty created the Earth as He saw fit. Since He obviously wished to place fishbones and shells in mines and inland cliffs, He must have had a reason for it. His will is not to be questioned.' As late as 1600 the Italian philosopher Giordano Bruno was burned at the stake for his heterodox views, which had included the theory that the seas had drowned the continents many times. Sir Walter Raleigh (1552–1618),on the basis of his own observations, came to the conclusion that the world must have been older than the Church generally allowed. (James Ussher, 1581–1656, Archbishop of Dublin and a scholar of Biblical chronology, worked out that the date of Creation was 4004 BC.) This taint of atheism was cited during his trial for treason in 1603 which later resulted in his execution.

Later in the seventeenth century, discussion about the history of the world and the interpretations of rock strata and fossils were not attended by such dire consequences. This was no doubt as a result of the flowering of the new scientific academies.

Meanwhile, the continent of North America was opening up and exciting new discoveries of bone deposits were being made. In 1706 Joseph Dudley, Governor of Massachusetts, obtained some large bones from the banks of the Hudson River which were considered to belong to an antediluvian giant.

In 1728 Sir Hans Sloane discussed the remains of tusks brought from Siberia. In Russian folklore these ivory tusks were thought to come from the *mammot* – a mythical creature that lived in the ground. But Sloane interpreted these remains as having belonged to elephants that had been overwhelmed in the Deluge.

In 1739 Baron de Longueuil, a senior French army officer discovered huge bones from the banks of the Ohio River. These were taken to Paris, where Baron Georges Cuvier (1769–1832), one of the founders of scientific palaeontology, compared them with elephant materials from Siberia and found them to be very similar. Subsequently further finds of giant bones, from what became known as Big Bone Lick in Ohio, were discovered by the great traveller John Bartram in 1762. This was followed up in 1765 by George Croghan, an amateur natural scientist, who reached the site and collected bones, but on the return journey was ambushed by Indians who killed most of his party. Croghan himself was wounded and subsequently ransomed. The following year, undaunted, he returned and succeeded in bringing out a large collection, part going to Benjamin Franklin (1706–90) of the American Philosophical Society in Philadelphia, the first scientific society in North America, and part to the English governor.

The English material, when it

▲ *The Exhumation of the Mastodon* by Charles Willson Peale records his great dig in 1802 for the giant skeleton in a flooded pit, which had to be baled out.

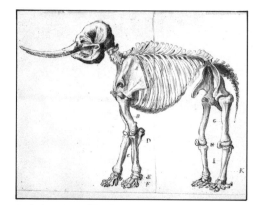

▲ A drawing of Peale's restoration of a complete mastodon skeleton by his son Rembrandt Peale.

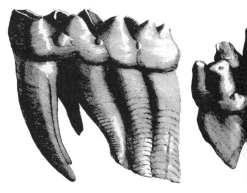

▲ Illustrations of mastodon teeth from Cuvier's *Recherches sur les Ossemens Fossiles* (1812). From the structure of the teeth,Peale

eached London, was described by one author as belonging to an elephant that fed on leaves, while surgeon William Hunter considered it was a type of carnivorous elephant, happily extinct, which he termed a 'pseudo-elephant or animal incognito'. The idea that the bones could be remains of organisms that were extinct was denounced by the clergy.

Thomas Jefferson (1743–1826), one of the first members of the American Philosophical Society and an early president of the United States, believed that fossils were not the remains of extinct animals, but from as yet undiscovered species. In 1799 he described the remains of the giant ground sloth, which he named *Megalonix* or 'great-claw', as a huge lion. When Jefferson visited Paris he was unable to convince the scientists that there were no extinct animals. Even his fellow members of the Philosophical Society, such as Judge George Turner, considered that such organisms were extinct, but that this was due entirely to the recent depredations of man.

The members of the American Philosophical Society were particularly excited by the discoveries of the remains of giant vertebrates. In 1799 a circular was widely distributed by the society signed by Jefferson, Caspar Wistar, Turner and Charles Willson Peale calling for news of finds of giant bones. Peale in particular had been actively on the lookout since 1784. As a result of the appeal, news came of further discoveries on a farm in New York State, in 1801. Peale bought the bones for 200 dollars and the right to dig for the rest of the skeleton for a further 100 dollars, plus a gun for the farmer's son and gowns for the daughter. His expedition, financed by the American Philosophical Society, returned to dig out the bones.

A giant skeleton was extracted, although unfortunately without the jaws. However, Peale continued his search and by probing the marshy ground with metal rods eventually located another skeleton – this time with the massive lower jaw. It became possible to construct a complete skeleton of this American mastodon.

The material was described in 1803 by Peale's son, Rembrandt, who considered that this was a creature *sui generis* (the only one of its kind) and was extinct. By this time he was able to refer to Cuvier who had described 23 extinct forms of animal life. Peale concluded, 'We are forced to submit to concurring facts as the voice of God – the bones exist, the animals do not.'

The members of the American Philosophical Society had developed concepts and a view of the world that was remarkably far ahead of their time. They felt free in the 1790s to question the literal interpretation of Genesis and to consider the extinct forms of life that had inhabited the Earth in the distant past. But such scientific activities were viewed with considerable suspicion. The pioneers of natural history in America were beset with opposition from evangelical sects which believed in the literal truth of the Bible and were deeply suspicious of any activity which might question the validity of their beliefs. Emotions on such issues are periodically stirred, as during the hounding of Prof. O. C. Marsh from the US Geological Survey (see pp 78–9), the Tennessee evolution trial in 1924 and the Creationist campaigns of the 1980s.

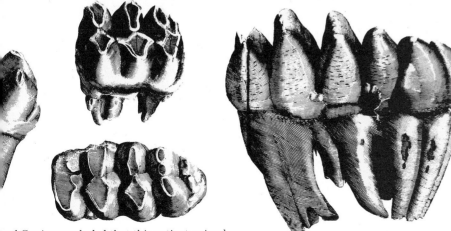

and Cuvier concluded that this extinct animal was a herbivore related to the modern elephant.

THE YEAR OF THE DINOSAUR

1822 marked the realization that recently discovered large petrified bones were of extinct giant reptiles.

In contrast to the United States, in Great Britain a completely different tradition emerged. The collection of fossils was considered an appropriate pursuit for members of both the aristocracy and the clergy. The British clergy had a long tradition of scientific curiosity, especially in the realm of natural history, where they collected everything from bugs and butterflies to fossils. Indeed the pioneers of palaeontology were two members of the established Church: the Reverends Conybeare and Buckland. Studying the wonders of God's creation was considered a worthwhile occupation for those who led leisured lives.

It was not until the end of the eighteenth century that interest in fossils quickened. Among the most enthusiastic of fossil collectors were a doctor, Gideon Mantell (1790–1852), and the Rev. William Buckland (1784–1856), the first Professor of Geology at Oxford University. Buckland was a devout Christian, but he recognized that fossils provided evidence for the existence of species of animals and plants, which had become extinct, preceding the existence of man. However, bearing in mind the Biblical account of the Creation, he was strongly opposed to the idea that one species evolved from another. James Parkinson, now best remembered for his description of Parkinson's disease, published *Oryctology: an Introduction to the Study of Fossil Organic Remains* in 1822 and remarked that the student 'will perceive decided traces of the vast changes which this planet has sustained and will see the remains of those beings with which it was inhabited previous to the creation of man'. He faced the problem that geological evidence seemed to contradict accepted Biblical teaching and suggested that a way round the problem was to consider the six days of creation, described in Genesis, as indefinite periods. Parkinson concluded that the changes he saw in the fossil record were incontrovertible evidence of the continual prescience and power of God. Nevertheless, this was a long way from a literal acceptance of the Scriptures. In spite of their sincere attempts to reconcile their new understanding and conclusions with the Bible, geologists and, indeed, geology itself were roundly condemned as ungodly from the pulpits of the land.

The evidence of the existence of former worlds did not really capture the public imagination until the discovery of bones which were recognized as having belonged to giant reptiles. William Buckland and his close friend, William Conybeare, spent a great deal of their time hunting for fossils. In 1821, Mary Anning found a complete skeleton of the marine reptile *Plesiosaurus* in Lyme Regis, England. Conybeare likened it to 'a serpent threaded through a turtle'; he also mentioned, in passing, the remains of a giant lizard from Stonesfield in Oxfordshire. These, part of a reptile jaw with large, serrated blade-like teeth, were in the possession of Buckland. The jaw had been examined by the famous French comparative

▲ Dr Mantell traced the *Iguanodon* teeth discovered by his wife to this quarry in Sussex, England. Here he found fossil bones in the sandstones of the Cretaceous period.

anatomist, Georges Cuvier, in 1818 and he constantly urged Buckland to publish an account of it. The first real mention of it in print was in Parkinson's book, where a tooth was illustrated with the as yet unknown animal from which it came, called *Megalosaurus*, meaning 'great lizard', a name invented by Conybeare. Parkinson stated that in life *Megalosaurus* would have been 12 m long and would have stood 2.5 m high.

In the same year, Dr Mantell, accompanied by his wife, visited a patient near Cuckfield just north of Lewes in Sussex, England. While he attended his patient, his wife went for a walk and, by chance, came across a number of large fossil teeth in a pile of rubble being used to surface the road. They were quite unlike anything she had seen among her husband's collections. The first mention of them was included in Mantell's book, *The*

Fossils of the South Downs: or, Illustrations of the Geology of Sussex, which appeared in May 1822, just a few months before *Megalosaurus* was announced to the world. The teeth were something of a puzzle. Their structure clearly indicated that they belonged to a large herbivore but remains of large, herbivorous mammals were unheard of in rocks of the Cretaceous period, 64–136 million years ago; the idea that they might have belonged to reptiles was inconceivable because no plant-eating reptiles were known. Mantell showed the specimens to his geological colleagues, but they were unable to offer any useful suggestions beyond the idea that they were fish teeth or belonged to modern mammals. In fact, Buckland warned him not to publish because he was not satisfied that they were the teeth of an ancient animal. Mantell then arranged for the geologist Charles Lyell to take one of the teeth to show the great Cuvier. Cuvier casually dismissed it as the upper incisor of a rhinoceros. Meanwhile, Mantell had traced the source of the rubble to a local quarry, where he found more material, including some foot bones. Cuvier identified these as hippopotamus. In spite of all this discouragement, Mantell was convinced that his material was extremely important and he persisted in trying to find out what it might be. On a visit to the Royal College of Surgeons he showed the precious teeth to the anatomist, Samuel Stutchbury, coincidently visiting at the same time, who said that, apart from their size, they looked exactly like the teeth of the Iguana lizard from the Americas. At Conybeare's suggestion, Mantell named the extinct reptile from which the teeth came, *Iguanodon* – Iguana tooth. In 1825, he presented to the Royal Society his evidence for the former existence of giant herbivorous reptiles – animals on which the giant flesh-eating *Megalosaurus* could have preyed. The previous year, Buckland had published the full description of his material of *Megalosaurus*, encouraged to do so by Cuvier's wish to include its description in his own book, *Recherches sur les Ossemens Fossiles*.

Iguanodon and *Megalosaurus*, together with the complete skeletons found by Mary Anning – the dolphin-like *Ichthyosaurus*, found in 1810, *Plesiosaurus*, 1821, and *Dimorphodon*, 1828 – presented the general public with an astonishing spectacle of a world before man inhabited by giant reptilian monsters. Based on the evidence of the fossil remains so far discovered, *Iguanodon* and

The quarrymen were accustomed to finding fossils in the rock, but did not notice the teeth.

Megalosaurus were seen as scaled-up versions of living reptiles, at which to marvel, but not fundamentally different from ordinary reptiles. At least they were not thought to be so, until the brilliant comparative anatomist Richard Owen (1804–92) calculated that, if the remains were scaled up with the proportions of living lizards, then some of them would have been over 60 m long. This was clearly absurd. He worked out from the vertebrae that their bodies would have been fairly short and, from looking at the way the limb girdles ('hips' and 'shoulders') and limb bones fitted together, he was able to demonstrate conclusively that these reptiles differed from all living reptiles in the way they stood and walked. Their limbs were held beneath the body in exactly the same manner as in modern mammals and they did not have the sprawling gait characteristic of modern reptile (the word 'reptile' is derived from the Latin verb *repere*, to crawl). Here then

were giant reptiles with the basic attributes of mammals in the way they moved. Owen insisted, with every justification, that these reptiles represented the crowning glory of reptilian life and he proposed that they should be grouped together in a new division of the reptiles for which he suggested the name *Dinosauria*, or 'terrible lizards' – a name that has stuck with them ever since.

After the Great Exhibition of 1851, the Crystal Palace, which had housed the pride of Britain's industrial achievements, was dismantled and re-erected in a large park at Sydenham in south London. At the suggestion of Prince Albert, life-size models of the dinosaurs were displayed in the same park. Models of *Iguanodon* and *Megalosaurus*, as well as other kinds of extinct reptiles, were specially made by the sculptor Waterhouse Hawkins under the supervision of Richard Owen. These concrete monsters remain on display today, even though the

Crystal Palace itself burned down in 1936. At the time the models were made, no complete or even partially complete skeletons of dinosaurs had been found anywhere in the world and, with the evidence Owen had to hand, his reconstructions were perfectly sensible and reasonable. There was no hint that his dramatic models would, within four years, have become curiosities, notorious for their astonishing inaccuracy. Nevertheless, Owen was fundamentally correct in his deduction of the way in which the limbs were held mammalian-like beneath the body.

The unexpected and dramatic change of opinion about the appearance of the dinosaurs came about with the first important discoveries of dinosaur remains in North America. In 1858 William Parker Foulke, a keen collector of curiosities, heard stories of large bones having been found 20 years earlier in New Jersey. With remarkable determination, he located

▲ The spiky thumb-bone of the *Iguanodon*. Originally one isolated example was found and all naturalists believed it belonged on the snout.

▲ *Iguanodon* under attack from *Megalosaurus* in a nineteenth-century print, drawn before a complete skeleton of *Iguanodon* had been found. The dinosaurs are reconstructed as

quadrupeds which is incorrect, but the way in which the limbs are held under the body, as in a mammal, is correct.

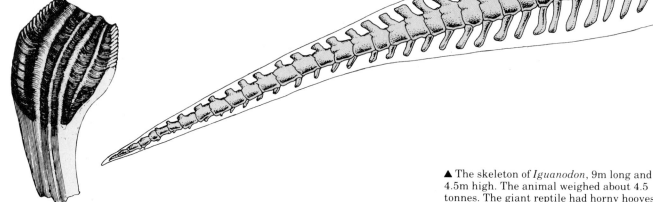

▲ An *Iguanodon* tooth. The large teeth with their worn surfaces were the first clue to the past existence of giant herbivorous reptiles.

▲ The skeleton of *Iguanodon*, 9m long and 4.5m high. The animal weighed about 4.5 tonnes. The giant reptile had horny hooves on its feet rather than claws. A complex of tendons in the tail show that it was held straight out as a balancing device when the animal walked.

the original quarry and re-excavated it, to be rewarded with the first complete dinosaur skeleton ever found. This remarkable specimen was described by Joseph Leidy (1823–91), the 'father of American palaeontology', and recognized as a relative of *Iguanodon*. He named it *Hadrosaurus foulkii* in honour of its discoverer. To everyone's amazement he found that the front-legs were shorter than the hind-legs by a third. The animal would have been 5m long and its hind-limbs would have raised it 1.8m off the ground but its fore-limbs only 1m. He surmised that its posture would have approached that of a kangaroo, the nearest parallel he could suggest. Later, the American palaeontologist, Edward Drinker Cope, discovered a skeleton of a flesh-eating dinosaur which he named *Laelaps* (in Greek mythology, Laelaps was a swift hunting dog that changed to stone in the very act of leaping). Cope's dinosaur had fore-limbs about a third of the length of the hind-limbs, which

proved conclusively that these carnivorous dinosaurs were two-footed, while Cope's evocative name was intended to emphasize the leaping nature of the dinosaurs. Since that time, studies of footprints have proved that in fact they waddled like birds (see pp 132–3). The idea of leaping dinosaurs then fell into disfavour until the 1960s, when advanced carnivorous dinosaurs, specialized for leaping, were discovered in North America.

The idea that *Iguanodon* itself habitually walked on its hind-legs was finally confirmed in 1878 with a discovery by Belgian miners. In the course of following a coal seam at Bernissart, they suddenly encountered sand full of gigantic bones. Apparently, a herd of *Iguanodon* had either fallen over the edge of a ravine or been washed into it, where their bodies were covered by sand and preserved. Louis Dollo of the Royal Natural History Museum, Brussels, set up a team of workmen and they extracted more

than 30 complete skeletons. Eleven mounted skeletons of this herd stand kangaroo-like in the museum. This discovery provided conclusive evidence of what a herbivorous dinosaur looked like. One of the oddest surprises was that the small bony spike found by Mantell and quite reasonably thought by him and Owen to have been a nasal horn, such as the living Iguanas possess, was actually a highly specialized thumb, a structure unique to *Iguanodon* among the dinosaurs. This thumb spike may have been used as a defensive weapon, but no one is certain of its purpose.

By the 1870s, throughout Europe and particularly in North America, active hunting for dinosaurs had begun in earnest. The notorious Cope-Marsh war broke out between two American palaeontologists, each determined to frustrate the efforts of his rival and be the first to discover and name the new marvels to be found as the West was opened up (see pp 78–9).

▲ This slab of rock, 2m across, contains fossilized *Iguanodon* bones. The haphazard arrangement of the bones highlights the problems encountered by Mantell in attempting to reconstruct *Iguanodon*.

THE COPE-MARSH WAR 1872–97

Fossil hunting in the American West was dominated by the aggressive rivalry between Cope and Marsh.

The major fossil collections during the latter part of the nineteenth century were made by two Americans, representative of the pioneering spirit of their age: Edward Drinker Cope (1840–97) and Othniel Charles Marsh (1831–99), who devoted their lives to the hunt for fossils as the American West was opening up. Even in the midst of the Indian wars they remained in the field.

Cope came from a wealthy Quaker family in Philadelphia. He was a precocious child and his interest in fossils began at the age of six when he visited the Peale Museum in his home town. His first scientific paper on 'The Primary Divisions of the Salamandridae . . .' was written when he was 19, after which he came under the wing of Joseph Leidy of the Academy of Natural Sciences. During 1863–4 he visited Europe and studied in Berlin where he met Marsh. Such was his brilliance that he was appointed Professor of Zoology at Haverford, Pennsylvania, at the age of 24. Within three years he had resigned and settled at his farm in order to devote himself to palaeontology.

The older Marsh came from Lockport, New York, and began his education later in life. His stolid hard work and the certificates he gained encouraged his uncle, the wealthy George Peabody (who later left him a fortune), to send him to university at Yale. It was not long before Peabody had been persuaded to provide Yale University with 150,000 dollars to establish the Peabody Museum of Natural History. In July 1866, Othniel Marsh was appointed America's first Professor of Palaeontology at Yale and put in charge of the Peabody Museum.

The US Geological Survey's pioneering expeditions surveying the American West found Cope and Marsh attached as palaeontologists to different expeditions. During the 1860s and 1870s there were four Geological Surveys. Cope was the official palaeontologist to the Hayden Survey and Marsh was associated with the surveys of Powell (who discovered the Grand Canyon) and King. The beginnings of their feud are attributed to Marsh pointing out that Cope had

reconstructed the marine reptile *Elasmosaurus* with the head on the tip of the tail, but in any case the differences in personality were so great that it needed little to provoke intense hostility between the two.

By 1872 Joseph Leidy was becoming increasingly concerned about the bitterness developing between Cope and Marsh and persuaded them at least to keep each other informed of what they were doing. In 1872 Cope and Leidy were excavating mammals from the Eocene of Fort Bridger, Wyoming, and Marsh was also excavating in the same area. One

member of the Cope-Leidy expedition was in fact on Marsh's payroll. When Leidy dutifully informed Marsh of his finds, this encouraged Marsh to get his team to work immediately in the same beds. When Cope uncovered important material he telegraphed his discovery to a newspaper without informing Marsh. But Marsh had already taken the precaution of bribing the telegrapher so that he was kept informed of Cope's progress. He then openly criticized Cope's method of identification. All this so disgusted Leidy that he withdrew from the bone feud to concentrate on writing.

▲ The Hayden Expedition field party in 1872, showing Dr F. V. Hayden seated at the far end of the table.

▲ A waggon and horses hauling fossil remains over rough terrain to the railway in Alberta, Canada, showing how difficult transport was for the early expeditions.

In 1876 Marsh hired an expert collector to go to Cope's sites. During 1877–80, vast quantities of material were shipped to Yale; subsequently Cope outbid Marsh so that during 1881–8 the majority of the finds went to Philadelphia. To give some idea of the amounts involved, the dinosaur fossil materials excavated by Marsh at Como Bluff amounted to about a ton per week for a decade.

In 1880 it was decided that the four US Geological Surveys should be combined. Marsh campaigned successfully for King against Hayden, and with Hayden out, Cope was cut off from his major outlet of publication. King's Survey was published in 1881 with a Supplement by Marsh, 'Odontornithes – a Monograph on the Extinct Toothed Birds of North America'. On receipt of this beautifully illustrated monograph, Charles Darwin wrote: 'Your work on these old birds and on the many fossil animals of North America has afforded the best support to the theory of evolution which has appeared within the last twenty years.' King resigned in 1881 to be replaced by Powell and Marsh became the US Geological Survey Palaeontologist.

From then on, Cope's fortunes went into decline. He had invested in Western mines which closed, and by 1888 he had become impoverished. Due to the intervention of Joseph Leidy in 1889, Cope was appointed Professor in the University of Pennsylvania, gave inspiring lectures and wrote a textbook because he thought the existing ones inadequate. Then in December he was ordered, according to a new rule instituted by Marsh, to hand over the collections he had made during the Hayden Survey, in spite of his having financed them himself to the tune of 75,000 dollars. Hayden had died in 1887, and without his support Cope took the only course open to him – the press. On 12 January 1890 the *New York Herald* splashed the feud across its pages under the headline: 'Scientists wage bitter warfare. Prof. Cope of the University of Pennsylvania brings serious charges against Director Powell and Prof. Marsh.'

Marsh replied by travelling to the University of Pennsylvania insinuating that 'poor Cope' might be cracking up and volunteering to find 'a more substantial scientist to replace Cope'. There ensued a pitched battle unparalleled in the annals of palaeontology but at the end of the day Cope continued teaching and excavating in Dakota and kept his specimens, which eventually he sold in 1895 to the American Museum of Natural History.

Both Cope and Marsh were firm Darwinists, but since Marsh was employed by the US government, he was answerable to Congress and the Fundamentalist lobby. When the 1893 budget for the US Geological Survey came before Congress in 1892, the Fundamentalists strongly urged their case, saying in the press, 'Birds with teeth – That's where your hard-earned money goes, folks, on some Professor's silly bird with teeth.' Louis Agassiz's son Alexander, at Harvard, joined in against the 'Godless', and Powell had to order Marsh's resignation. His great work that had so pleased Charles Darwin enabled the Fundamentalists to bring him down.

Marsh had used up his entire fortune and so had to depend on a salary from Yale. He died in 1899 with less than 100 dollars in the bank. Cope continued as Professor at Pennsylvania until his death in 1897. The Cope-Marsh war thus ended. Marsh had in fact collected more dinosaurs than Cope; he had described 19 genera to Cope's 9. This was due, however, to Cope's wider interest in all types of fossils. In any case, the American nation inherited the most prodigious fossil collection ever amassed.

▲ This drawing was made in 1873 under Cope's direction and illustrates his incorrect conception of uintatheres as elephants.

▲ Cope and Marsh competed with each other in collecting fossil remains of the early mammals, the uintatheres. They interpreted the fossils in different ways. This drawing shows the generally accepted view of the uintathere which agrees with Marsh's interpretation.

CHARLES DARWIN AND NATURAL SELECTION

**From his observation of giant tortoises and finches in the
Galapagos Islands, Darwin was convinced that species could change;
how and why they did so was explained by natural selection.**

The realization that fossils are a historical record, albeit imperfect, of the living world can be attributed to the work of Charles Robert Darwin (1809–82). The fossil record indicates that living organisms have changed over long periods of time and this was the basis of Darwin's theory of evolution.

When Charles Darwin left on his epic voyage round the world in HMS *Beagle*, 1831–6, as Captain Fitzroy's companion and gentleman naturalist, like most of his contemporaries he was convinced of the immutability of species. However, the wealth of observation which he made during these years helped change his mind. On the plains of Patagonia he found fossils of extinct animals related to the living tree-dwelling sloths and the armadillos of South America. This showed him there were animals, now extinct, which had been succeeded by similar kinds of animals. He did not, however, at this point see that the one might have changed into the other, he just recognized that they were related.

It was during his short stay in the Galapagos Islands that his first doubts arose about the idea of the immutability of species. The vice-governor of the islands pointed out that he could recognize the island of origin of the giant land tortoises by differences in the patterns of their shells. It was Darwin's observations of the small finches of the islands, all similar in appearance but with strikingly different feeding habits and beaks, that really convinced him. He saw that their existence could be explained if it were admitted that species could change.

The observations Darwin made during the voyage of the *Beagle* presented overwhelming evidence of change occurring through time. Charles Lyell's book, *Principles of Geology*, that Darwin took with him, demonstrated that the geological history of the Earth was one of change. Why should such an approach not be equally appropriate to the living world?

Lyell's principle of 'uniformitarianism' (see pp 10–11), whereby the distant past can be explained by processes taking place in the present working over a long period of time, provided Darwin with the vital clue. Darwin, contrary to many people's belief, did not invent, nor was

he the first to introduce, the idea of evolution – others, such as his grandfather Erasmus Darwin, had discussed the issues, but they had been unable to present sufficient evidence to carry conviction. This was Charles Darwin's monumental contribution to human thought. He presented the idea of evolution to the world based solidly on observation and logical deduction.

Darwin's theory of evolution was

in fact two separate theories: the first Darwin described as the theory of descent with slow modification – a historical view based on change through time; the second stated that the process by which change occurred was natural selection. Darwin began his book *On the Origin of Species by means of Natural Selection or the Preservation of favoured races in the struggle for life* with chapter 'I: Variation

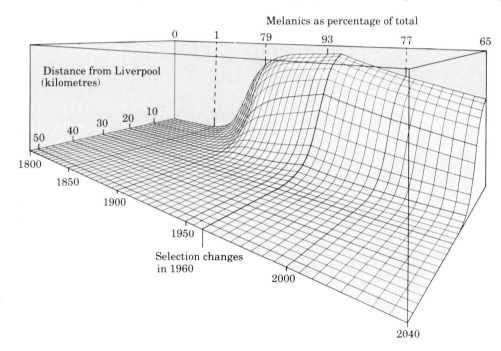

▲ A computer model showing how the proportion of melanic peppered moths fluctuates as a result of natural selection over a period of 200 years in an area extending from the industrial city of Liverpool, England.

▲ Both the dark and light-coloured moths can be conspicuous depending on their environment; they will then be easily spotted by birds.

under domestication', demonstrating how animal and plant breeders select for the desired features. The book went on to deal with chapters 'II: Variation under nature', 'III: The struggle for existence' and 'IV: Natural selection'.

The idea was simple enough. The examination of any animal or plant species illustrates that the individuals vary among themselves. It is also common knowledge that animals and plants produce more offspring than ever reach maturity. Just as animal and plant breeders deliberately weed out the unfit, so in the natural world the environment acts as a kind of filter – the individuals that survive are likely to be those that are better adapted to the environment. The selection is natural.

Variation in a population was the starting point for Darwin, although he was unable to account for its origin as the science of genetics was still far into the future. But since populations were varied, natural selection provided a convincing explanation of why species should change gradually over generations.

An important confirmation of the role of natural selection in evolution was noted after Darwin's death. This was the change in the population of the peppered moth *Biston betularia* in the industrial cities of Britain. It was first recorded in 1886 and was discussed in 1906 by the Evolution Committee of the Royal Society. The dark or melanic forms were conspicuous in the countryside and hence immediately devoured by birds; however, in the smoky grimy cities they were effectively camouflaged. Within a few years the population in the industrial towns consisted almost entirely of dark moths. With the change in the atmosphere following the Clean Air Act of 1952, the proportion of melanic forms began to fall. The studies on the peppered moth in the 1950s established that natural selection takes place. This example is by far the most obvious, but it is dealing with a single characteristic – colour. The situation is usually more complicated.

Detailed field studies have been made by Prof. A. J. Cain of the snail *Cepaea* and a wide variation in shell patterns has been found, both in shell colour and in the type of spiral banding. As the generations have succeeded one another, this variety has been maintained. The snails are eaten by thrushes which smash them on stones, and thrush middens show the type of selection of snails that has taken place. The snails inhabit different environments: in open grassland banded snails are conspicuous and get picked out by the thrushes, whereas in the undergrowth the unbanded shells are more easily found. In addition to this there are seasonal changes: in the spring and summer yellow shells are better camouflaged, whereas in autumn and winter brown and pink shells stand a better chance of survival. This means that snails have advantages and disadvantages at different times of the year and as they wander from place to place. The effect is to maintain a species that remains variable and is thus adapted for life in a changing environment.

Furthermore, Darwin presented evidence to show that the physical separation of varieties can lead to their eventual separation as different species. Even the present-day Creationists concede Darwin's thesis on the origin of, for example, the different Galapagos finch species and the mechanism of natural selection. But they claim that such changes do not account for the origin of what they term 'kinds', such as mankind, ape-, dog-, cat- and horse-kind (this system of dividing living things is not precisely defined by the Creationists); these 'kinds' are believed by the Creationists to have been originally created by God. Darwinians argue that with the evidence of natural selection, the origin of species and the enormous length of geological time, the only rational explanation for the diversity of life on Earth must be evolution.

▲ There is enormous variety of colour and pattern in the banded snail, *Cepaea*, which is a classic example of a polymorphic species, i.e. one which has various quite different external forms.

EVOLUTION AND THE FOSSIL RECORD

The fossil record rarely shows all the stages in the evolution of a species. This has given rise to an alternative view to Darwin's gradualism whereby changes occur suddenly.

The process of natural selection that can be observed involves very small changes, since great changes take place over a long timespan. As a result, the final proof of evolution can only be found in the fossil record. This much is agreed by all sides. One of the problems which Charles Darwin faced was the gaps in the fossil record: 'Why then is not every geological formation and every stratum full of such intermediate links? Geology assuredly does not reveal any such finely graduated organic chain and this, perhaps, is the most obvious and gravest objection which can be urged against my theory. The explanation lies, as I believe, in the extreme imperfection of the geological record.' So wrote Darwin in *The Origin of Species* (1859). Later he continued: 'The abrupt manner in which whole groups of species suddenly appear in certain formations has been urged by several palaeontologists, and none more forcibly than by Professor Sedgwick, as a fatal objection to the belief in the transmutation of species. If numerous species, belonging to the same genera or families, have really started into life all at once, the fact would be fatal to the theory of descent with slow modification through natural selection.'

In the nineteenth century the ecclesiastical geologists of Britain, as well as the great Baron Georges Cuvier of France, firmly upheld the view that the fossil record documented a series of catastrophes. But Darwin continually emphasized the old naturalist's adage *Natura non facit saltum* – Nature does not make jumps. Curiously, even Thomas Henry Huxley (1825–95), who proselytized Darwin's

ideas, took the saltationist view of change via sudden jumps, which he noted he held 'much to Mr Darwin's disgust'.

The problem of gaps in the fossil record has led in recent years to a lively debate among palaeontologists. Many have concluded that the situation can best be explained by short periods of evolutionary bursts alternating with periods of stasis or equilibrium. This theory of 'punctuated equilibria' was proposed as a direct alternative to Darwin's gradualist view. It has been forcibly argued by

the early Marxists, and the contemporary Harvard professor Stephen Jay Gould, that Darwin's gradualist view of change was opposed to sudden revolutionary jumps. Gould asked the question, 'Why were Lyell and Darwin such convinced gradualists if, in fact, they did not see it in the rocks? I think there are lots of reasons. I think it is pretty undeniable that one important source of gradualism, if not the most important, really has to do with political ideology.' Gould insists that Darwin was reflecting the basic mid-Victorian capitalist ethic. As Karl

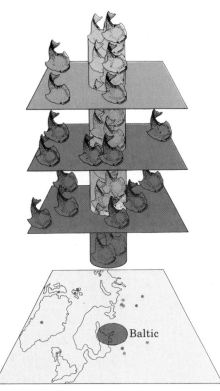

MAN·IS·BVT·A·WORM·

▲ *Punch's* view of Charles Darwin in 1882, the year of his death.

◀ The map shows the distribution of ostracoderms in the Middle and Upper Devonian. The diagram representing the fossil record of ostracoderms shows their gradual evolution in the Baltic by the means of the central column, and three major periods of migration from the Baltic, which give the impression of evolutionary jumps.

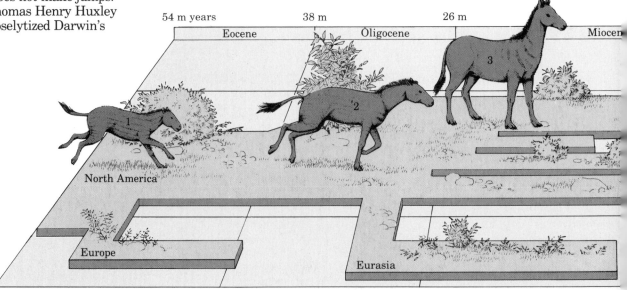

Marx wrote in a letter to Friedrich Engels in 1862, 'It is remarkable how Darwin recognizes among beasts and plants his English society, with its division of labour, competition, opening up of new markets, "invention", and the Malthusian "struggle for existence" '. Gould has recently been promoting the concept that the fossil record appears jerky simply because such is the process of evolution. He has said that this idea is, 'replacing the concepts of gradualism with the flip-like style of change which has always been appreciated within Marxist philosophy for a long time'.

There is now a concerted assault on Darwin's concept of gradual change through time. It is true that the concept of sudden jumps would explain the apparent jerky nature of the fossil record and these ideas are popular probably because of this. However, if general conclusions and speculations about the fossil record are to be drawn, the first task should be to examine all the available evidence. For example, when fossils of ostracoderms, early jawless armoured vertebrates, are examined throughout the world they are found in sequences of rocks characterized by the remains of the same types of fossil, but the later rocks contain more advanced forms. This gives the impression of sudden jumps in the fossil record.

However, this situation can be explained. The evolution of the ostracoderms took place in a major faunal province that was situated 400 million years ago in what is now western Russia, the Baltic Province. Evolution was a slow gradual affair and at particular times there were waves of migration from the evolutionary centre to outlying provinces, where the immigrants became established but did not undergo any further evolutionary change. However, they were subsequently replaced by succeeding waves of new immigrants. The fossil record, therefore, shows slow gradual change at the centre but a series of jumps in the outlying areas as one group of immigrants replaced another. This gives a spurious impression of evolutionary jumps.

Whenever a reasonably complete fossil sequence is studied, the evidence points to gradual change. The classic story of horse evolution first worked out in Europe illustrates this. Four major stages of horse evolution were documented, from the small dog-sized *Hyracotherium*, through three-toed forms, culminating in the modern horse *Equus*. Unfortunately there were no connecting links. These had to await the discoveries in North America. There, the full sequence of horse evolution was preserved; it proved to be gradual with a series of migrations across the Bering Land Bridge to Eurasia.

Recent studies on Radiolaria, microscopic unicellular animals preserved in deep-sea sediments in the Pacific, Antarctic and Indian Oceans, which show continual sedimentation over millions of years, demonstrate an unbroken fossil record; all the changes that are documented are gradual.

Finally, Gould has claimed the record of fossil man as evidence of sudden jumps, but in 1981 a detailed re-examination of the recent new discoveries shows that they conform to a gradualist pattern (see pp 162–3).

There is no doubt that ideologies can colour some scientists' interpretations, but in the end it is the data that must determine the outcome of the debate. The objections to gradual evolution by both Creationists and punctuated equilibrists rely heavily on gaps in the fossil record, an insecure base for any theory.

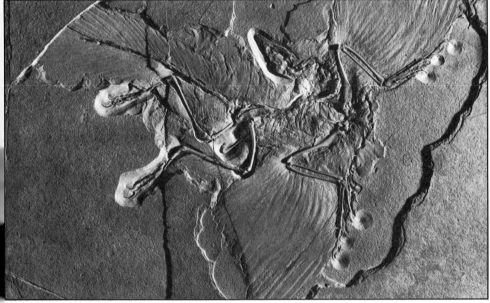

▲ The Berlin specimen of *Archaeopteryx*. The fossil was discovered a few years after the publication of *The Origin of Species* and represents an intermediate stage between reptiles and birds. The skull, teeth, front claws and tail are reptilian but *Archaeopteryx* also has bird-like feathers.

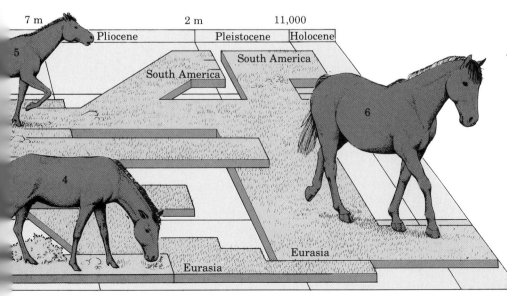

◄ The main evolution of the horse took place in North America, although there were several migrations into Eurasia and one to South America; but during the Quaternary horses became extinct in America. The major evolutionary change from browsing to grazing occurred with the spread of grasslands in the Miocene.

1 *Hyracotherium*
2 *Mesohippus*
3 *Merychippus*
4 *Hypohippus*
5 *Pliohippus*
6 *Equus*

CHAPTER 4

TRACING THE ORIGIN AND EVOLUTION OF LIFE

The origin of life and its early evolutionary
stages cannot be understood through processes which
occur today. Conditions on the primitive Earth were
very different particularly as the Earth's atmosphere
must have been free from oxygen.
We have to work out what the Earth was like then
by means of chemistry and biochemistry
as well as examining the geological and fossil record.
The further back we go, the scarcer is the fossil evidence.
However, there are some primitive organisms living
today that have existed for about 3,000 million years.
This kind of fossil evidence lends the
perspective of time to our understanding of
the slow process of evolution.

THE ORIGIN OF LIFE

**The origin of life is not known, but attempts have
been made to understand the composition of the Earth's primitive
atmosphere and how life could have formed.**

Among the most ancient sedimentary rocks of the Earth's crust are iron ore deposits of tens of thousands of millions of tons which are the major sources of iron ore in the world. These iron deposits, known since the last century, comprise thin iron-rich bands with thin silica-rich bands in between, hence the name 'banded ironstone formation', or BIF, given to them. All attempts to interpret these rocks on the basis of processes that can be directly observed at the present day have singularly failed. This should occasion no real surprise as it is extremely unlikely that conditions at the beginning of the Earth's history were comparable to what is observed today. Indeed, life and the environment have gradually changed together.

There are two fundamentally different approaches to studying the origin of life: the first explores the possibility that one-celled organisms evolved from complex organic molecules that were produced in an atmosphere devoid of oxygen. The other view states that the origin of the simplest form of life is so complex that there is no way it can be explained as the result of random changes. In 1981 the astronomers Fred Hoyle and N. C. Wickramasinghe in their book *Evolution from Space* took this stand and concluded that life must have been seeded onto the Earth from space. However, there is nothing that can justify not attempting a rational scientific explanation.

The very first problem in trying to explain the origin of life is the complexity of the chemical structure of living things. How is it possible that complex organic molecules, the building blocks of life, were synthesized from simple gases in the Earth's primitive atmosphere? We need to try to work out what the composition of the atmosphere was likely to have been. There are two ways of tackling this question; one is to study the atmospheres of our nearest neighbours, Mars and Venus, which are dominated by carbon dioxide; there is abundant water-ice on Mars, and water vapour on Venus. The second is to examine the composition of the gases that come out of volcanoes. It is likely that the oceans and atmosphere of the Earth were both the product of the initial stages of the formation of the Earth, when the lighter material escaped from the interior in the process known as 'outgassing'. Modern volcanoes in action show the same process. From both these studies it is reasonable to infer that the major constituents of the early atmosphere comprised carbon dioxide and water with nitrogen, hydrogen sulphide, ammonia and methane, all of which are volcanic gases. What is generally agreed is that there was no free oxygen.

In 1953 Stanley L. Miller, an American research student, devised a simple experiment in which a mixture of gases believed to have been present in the primitive atmosphere, were heated and subjected to an electrical spark. Miller discovered that complex molecules like amino acids, which make up proteins, were synthesized. There was no need to invoke an outside force beyond an electrical

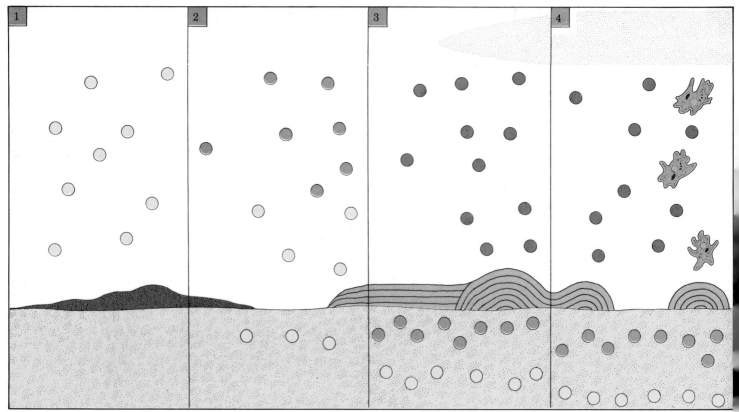

▲ **The origin of life**
1 The earliest forms of life were heterotrophic bacteria that lived by breaking down molecules in oxygen-free (anoxic) conditions. 2 Bacteria capable of photosynthesis existed together with the primitive bacteria, but still in anoxic conditions. 3 Cyanobacteria (blue-green algae) produced oxygen as a result of photosynthesis. This combined with dissolved ferrous iron to form ferric iron oxide. Banded ironstone comprised alternate layers of iron oxide and silica with some limestone. 4 As iron was being used up in the oceans, only limestone was precipitated and stromatolites were produced. Oxygen escaped into the atmosphere and some of it contributed to the formation of the ozone layer.

charge. Subsequently another American, Sidney W. Fox, was able to demonstrate that with dry heat, such as would be encountered in association with volcanic activity, mixtures of amino acids would join up together to produce chains of protein-like materials, called 'proteinoids'. One of the most remarkable features of proteinoids is that if they are quenched in water they aggregate spontaneously into minute spheres and develop a boundary layer which resembles a cell membrane. If placed in a suitable medium, such as salty water, proteinoids will grow by incorporating dissolved proteinoid and divide in a manner reminiscent of bacteria. Even more surprising is that such proteinoid globules can catalyze the decomposition of sugars.

These experiments do not create life, but they show in a convincing manner that many of the properties of living things are the consequence of their chemical make-up. It is acceptable to postulate the synthesis of complex molecules during the activities associated with the Earth's outgassing phase together with electrical storms and ultraviolet radiation. Complex organic molecules are known to exist in meteorites – the carbonaceous chondrites – and it seems as if such synthesis is not unusual in the universe. We know that similar molecules exist in interstellar space.

From these experiments it can be postulated that the first forms of life comprised minute spheroidal organisms, primitive heterotrophic bacteria which maintained themselves by breaking down the organic molecules that were being formed. This is a type of fermentation and must have taken place in anoxic, or oxygen-free, conditions. The most primitive of living bacteria are unable to survive in the presence of oxygen. The primitive bacteria that are capable of photosyn – thesis cannot photosynthesize in the presence of oxygen. Nor is oxygen a product of their photosynthesis. The sulphur bacteria, found in hot volcanic springs, for example, act on hydrogen sulphide which they split for the hydrogen, which is then used with carbon dioxide to synthesize sugars. The fact that the most primitive of living bacteria are poisoned by oxygen suggests that they must have originated at a time when free oxygen did not exist on the Earth. There is some geological evidence that confirms this. The mineral uraninite (UO_2) was formed on ancient stream-beds, and in the presence of oxygen this mineral is converted to U_3O_8 which is soluble and hence dissolves in water. The presence of uraninite indicates the absence of any oxygen.

The ability of bacteria to photosynthesize, using water instead of hydrogen sulphide, marked perhaps the most significant step forward in the history of life on the planet. This took place over 2,000 million years ago and the waste product of the process was oxygen.

The effect of the release of oxygen was dramatic: iron in the ferrous form was dissolved in the oceans and in the presence of oxygen changed to ferric oxide, which then came out of solution and was precipitated onto the sea-floor. The world's oceans rusted. The banded ironstone formations began to form. The banding of thin iron-rich layers interspersed with silica-rich layers containing calcium carbonate shows that the rusting process was not an even one but was probably seasonal. During spring and summer, the cyanobacteria, or blue-green algae, flourished releasing oxygen which resulted in the production of ferric oxides and their precipitation. When their photosynthetic activity died down in the autumn and winter, the fermentation by the anaerobic bacteria in the bottom sediments would release carbon dioxide which would lead to the formation and precipitation of calcium carbonate. There seems to be no other explanation to account for BIF. From a study of primitive living organisms and their chemistry, a unique part of the geological record has been elucidated.

▲ Electron micrograph of the oldest known bacterium from Precambrian rocks in South Africa it;is about 3,000 million years old. Bacteria like this provide evidence of the earliest stages of evolution.

▶ A piece of banded ironstone from Africa.

STROMATOLITES

The earliest known fossils, stromatolites, were formed from calcium carbonate. They are still being formed today, having existed for nearly 3,000 million years.

One of the major difficulties that was used to confront evolutionists was the absence of Precambrian fossils – the apparent burst of the fossil record at the beginning of the Cambrian was taken as conclusive evidence that the animal world had not gradually evolved but had suddenly appeared.

Although it was not recognized until this century, there is fossil evidence throughout most of the Precambrian, from about 3,800 million years ago up to the beginning of the fossiliferous Cambrian 570 million years ago. The earliest fossils occur with the first banded ironstone formations recently found in Australia. These common fossils are known as 'stromatolites' and are thin layers of limestone which may be in the form of pillars or mounds. The name comes from the Greek *stroma*, a bed, and means 'stone blankets'. These structures were first discovered at the beginning of this century in North America by Charles D. Walcott (1850–1927), who interpreted them as fossil reefs formed by a type of algae. Most geologists dismissed this interpretation and believed that they were simply the result of chemical precipitates that had nothing to do with living things. In 1954 fossil blue-green algae, now termed 'cyanobacteria', and other bacteria were recognized in these deposits and Walcott's original interpretation gradually came to be accepted by all scientists working in this field. Later, in the 1950s, blue-green algae producing stromatolites were found in Australia, providing the conclusive evidence needed to settle the issue. This discovery was made in Shark Bay, an excessively salty lagoon; it is an environment too hostile for normal marine organisms so that the primitive cyanobacteria and other bacteria were able to develop without any undue interference. This discovery in the 1950s of living structures best known from rocks 3,000 million years old came as a considerable surprise.

The heaped-pancake appearance of the stromatolites is produced by successive depositions of layers of calcium carbonate and, as with the banded ironstone formation, the stromatolites are the product of blue-green algae and anaerobic bacteria. The process is similar to the formation of banded ironstone, but without the iron-rich layers. This is because, where great sheets of cyanobacteria occurred in the shallows, the water had been cleared of iron by the oxygen they produced. The carbon dioxide given off when the cyanobacteria were not photosynthesizing resulted in the precipitation of calcium carbonate alone. Vast quantities of Precambrian limestones were formed in this way. It

▲ These are living stromatolites at Hamelin Pool, Western Australia – a hypersaline environment hostile to most organisms which allows cyanobacteria to flourish.

▲ Precambrian stromatolites in Canada; erosion has exposed their concentric layers.

was the oxygen production of the cyanobacteria that finally cleared the world's oceans of iron. This in turn caused the deposition of banded ironstone formation to come to an end about 1,500 million years ago, and this type of rock was never again formed.

At this stage the stromatolites were flourishing and continued to produce oxygen, but there was no longer a source of ferrous iron to take it up so that some of it was released to the atmosphere. The abundance of free dissolved oxygen in water resulted in the anaerobic bacteria being completely restricted to anoxic environments. In fact, such environments within sediments and in hot volcanic springs represent some of the original conditions in which life must have first developed on the Earth. The release of free oxygen to the atmosphere had another consequence, which was to become significant in the distant future: an ozone layer was formed in the atmosphere that had the effect of reducing the amount of ultraviolet radiation reaching the Earth's surface, since ozone acts as a barrier to the passage of ultraviolet radiation.

As the accumulation of oxygen continued and the era of the stromatolites lasted for some 1,000 million years more, the survival of organisms in the water involved the ability to tolerate the presence of oxygen in the environment. The decline of the stromatolites was associated with the rise of more efficient organisms capable of actively using the waste product, oxygen. As the amount of oxygen built up, so the ozone layer became more effective. This caused a complete change in the type of radiation reaching the surface of the Earth. An abundance of oxygen gave organisms a new opportunity: the burning up of complex molecules to release energy is best effected by oxygen. To use oxygen in the process of respiration as opposed to, say, fermentation is exceedingly efficient. The ability to use the energy of sunlight to combine carbon dioxide and water to form sugars, a process that gives off oxygen as a waste product, and then to use the oxygen to break down sugars into carbon dioxide and water, thus releasing energy, was a fundamental advance. The energy supply for the construction of complex molecules was now simply sunlight.

In considering the origin of life, it is important to realize that the early conditions on Earth can be replicated in the laboratory and many of the first steps towards the synthesis of life have already been achieved. This progress would not have been accomplished had the scientific community accepted the notion that life and its origin was something beyond man's knowledge and simply evidence of the existence of a Creator. It was an area of pure speculation, but experiments starting in the 1950s have elucidated the types of chemical reactions which most probably took place. These, in conjunction with the fossil record and an analysis of the early types of rock, have shown in broad outline the major developments of the first few thousand million years of Earth's history.

▲ Fossil algal stromatolites at Enorama Creek, Flinders Ranges, Australia.

THE FIRST PLANT AND ANIMAL CELLS

The release of oxygen into the atmosphere coincided with a momentous change in the living world — the development of the first plant and animal cells.

The increase of oxygen as the result of the photosynthetic activity of cyanobacteria led first to the rusting of the oceans. Oxygen combined chemically with ferrous iron to form ferric oxide, which was precipitated onto the sea-floors. Once the oceans had been swept clean of iron in this way, the oxygen came to accumulate in the atmosphere and it was the turn of the land to be rusted.

This was caused by the iron minerals that were deposited. Iron minerals derived from igneous rocks are generally dissolved, but in the presence of oxygen will be deposited as ferric oxide around sand grains or in the spaces between them. Ferric oxide is characterized by its red colour. About 1,500 million years ago the first red beds made their appearance in the geological record; the banded ironstone formations were no longer being formed. A fundamental change in the atmosphere and its chemical composition had taken place and this had been effected through the agency of living things. Indeed, all the oxygen that living things breathe today is the direct product of plant life.

From this period onwards, the geological record can be effectively understood in terms of uniformitarianism. This means that, apart from relatively minor fluctuations, the basic processes that can be observed at the present day provide a sufficient explanation of what is preserved in the geological record.

The fossil record at around 1,450 million years ago begins to reveal evidence of a major, perhaps the most revolutionary, change in the history of life. Up until this time all the fossils were microbes, never larger than about 60 micrometres, notwithstanding the fact that they may have lived in large mat-like aggregations. They were, without exception, comparable in their dimensions to bacteria and blue-green algae. These organisms are characterized by not having their genetic material, that is, the deoxyribonucleic acid (DNA), enclosed in a nuclear envelope within the cell. It tends to be organized in the form of a single long strand. These types of organism reproduce simply by splitting. It has become recognized since the 1960s that the major division in the living world is between these types of

primitive organism known as 'prokaryotes' – and all the rest, known as 'eukaryotes' – which include unicellular animals and plants, moulds, fungi and multicellular animals and plants. The term prokaryote is derived from the Greek *pro*, first, and *karyon*, kernel.

The eukaryotes, or true karyotes, differ from the prokaryotes by having their genetic material in the form of

chromosomes which are enclosed in a membrane in a nucleus within the cell. They reproduce by cell division through the complicated process of mitosis. There is a further method of division, meiosis, where the genetic material divides so that each resulting daughter cell has only half the genetic material. This cell then combines with a similar cell from another individual. By means of this sexual reproduction

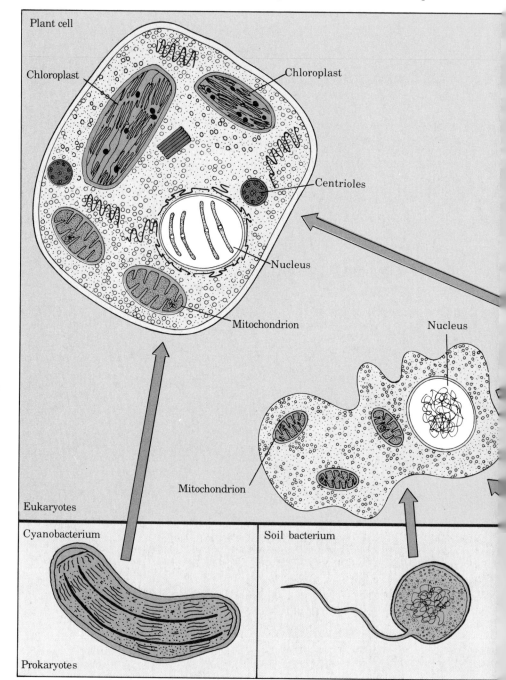

The origin of plant and animal cells.
The diagram shows that the first eukaryotic animal cells may have formed when prokaryote bacteria were ingested by another organism to become symbionts within the cell. The mitochondria, cilia, flagella and centrioles may have all originated from

the genetic material of the species is continually being mixed, which produces variety in the organisms; the combinations of characteristics which result can be quite different from the parent forms. Because of these variations, the process of natural selection, or survival of the fittest, comes into play. One of the interesting features of eukaryote cell division is that it can only take place in the presence of oxygen and it is assumed, therefore, that this method did not evolve until sufficient free oxygen was available in the atmosphere and oceans.

The first hint as to the existence of eukaryotes is in the size of microscopic fossils that begin to be found in rocks 1,450 million years old.

Some of the earliest of these 'larger' cells were about 100 micrometres in diameter, but larger fossils with diameters of 600 micrometres are found in Precambrian rocks of North America, Siberia, China, Africa and Australia. This evidence suggests that the eukaryotes were in existence, as all known prokaryotes are only about 60 micrometres in diameter.

Unfortunately, the organization of the genetic material and the method of reproduction in these two cells are so different that it is difficult to imagine how eukaryotes could have evolved directly from prokaryotes. Eukaryote cells are larger, but this in itself need not be significant. However, the cells differ in one important respect: the

interiors of the eukaryote cells, unlike prokaryotes, are highly organized into small organs or 'organelles'. One of the most interesting of these is the 'mitochondrion'. This is commonly known as the power-house of the cell and is concerned with energy exchange. The curious feature of the mitochondrion is that it has its own genetic material. Moreover, the mitochondria reproduce independently of the rest of the cell, although during normal cell division some mitochondria will go with one daughter cell and some with the other. The genetic material, DNA, of the mitochondria is organized in exactly the same way as that of prokaryote aerobic bacteria. All the available evidence leads to the conclusion that mitochondria were originally prokaryote bacteria that became part of the eukaryotic cells through symbiosis. It has been suggested that such bacteria were ingested by another organism and, instead of being destroyed, were maintained as they were able to deal effectively with the new poison, oxygen. This theory, which was first put forward by Professor Lynn Margulis in 1971, has only recently become generally accepted.

The eukaryotes have an organelle called the centriole, which plays an important part in mitosis. In eukaryotes there are also cilia and flagella which have the same basic structure as the centrioles. They contain their own DNA similar to the arrangement of the thread-like prokaryote and, like the mitochondria, seem to have once been independent organisms which became incorporated into the eukaryote cell – probably spirochaete bacteria, which they resemble. These organelles are common to all eukaryotes, both animal and plant. Green plants also have another type of organelle concerned with photosynthesis, the chloroplasts, which contain chlorophyll. These too have their own DNA and reproduce independently. In fact they are cyanobacteria ingested as symbionts (organisms of different kinds which associate to their mutual advantage).

It seems reasonable to conclude that the eukaryote animal cell evolved first and it was the subsequent acquisition of internal cyanobacteria that gave rise to the first true plants. The study of both prokaryote and eukaryote cells and their geological history leads to the astonishing conclusion that the simplest animal and plant cells are the result of the combination and active co-operation of several different types of primitive living organism.

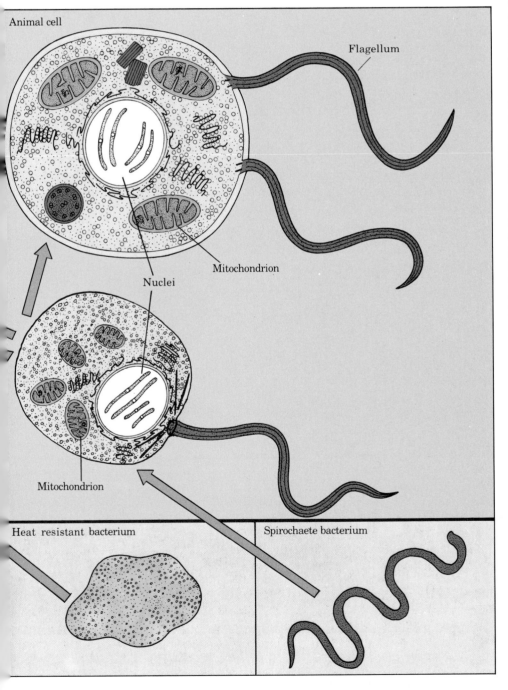

Animal cell

Flagellum

Mitochondrion

Nuclei

Mitochondrion

Heat resistant bacterium

Spirochaete bacterium

independent bacteria. Plant cells may then have been formed when photosynthesizing bacteria, or blue-green algae, were

incorporated; these subsequently became chloroplasts which are responsible for photosynthesis in plants.

PRECAMBRIAN ANIMAL LIFE

Traces of life from Precambrian times are rare, preserved only under exceptional circumstances. Fossil evidence shows that animal life 680 million years ago consisted of soft-bodied sea creatures.

The fossil record before the Precambrian era was blank until recent times. During this century, however, fossil finds have shown that although it was not greatly varied, animal life certainly did exist in the Precambrian.

One of the first important Precambrian fossils to be found was described in 1927 from an erratic boulder of sandstone in Sweden. It comprised the mould of an organism that seemed to have had a body with paired appendages with tiny segments. This fossil was named *Xenusion* and was interpreted as an onychophoran, an animal halfway between an annelid, or segmented worm, and a joint-limbed arthropod. *Xenusion* became a famous fossil and was featured in all popular books on the history of life, as an animal illustrating the intermediate stage between two major divisions of the animal kingdom. Before the First World War a number of similar fossils were discovered in

south-west Africa (Namibia) and these were described in the 1930s, but they were not linked to *Xenusion*. During the 1940s major discoveries of fossil jellyfish were made in the Ediacara Hills near Adelaide in South Australia. Although these fossils were first thought to be early Cambrian, a detailed study of the geology established that they came from well below the oldest Cambrian rocks and so were Precambrian in age.

The preservation of fossils such as jellyfish seems at first to be unlikely, but when these animals are washed up onto sandy beaches they are sufficiently firm to take the weight of a human being. If they are covered sufficiently rapidly they stand a chance of leaving a permanent imprint in the sediment. Detailed systematic collecting of fossils from Ediacara resulted in a large fauna being described in the 1970s. The fauna consisted of moulds and casts in the fine-grained sands on the under-surface

of the sediments that buried them. The method of collecting these fossils is to turn over slabs of rock on site so that the under-surface is exposed, then leave them until the elements have washed the fossils clean. The overturned slabs can then be re-examined and all the worthwhile specimens collected. By far the most dominant of the fossils were some 19 different types of jellyfish. Since floating forms are the most likely to be washed up onto beaches, it does not necessarily mean that the life in the seas was dominated by jellyfish. Some forms were related to the Portuguese man-of-war which is not a true jellyfish, but colonies of related organisms with gas-filled floats.

Jellyfish are among the more primitive of the multicellular animals, the coelenterates, which are made up of two layers of cells and catch their food by means of tentacles which bear poisonous stinging cells. The evident success of coelenterates in the oceans

▲ The main fossil site of Precambrian rocks at Ediacara, near Adelaide, S. Australia.

▲ A fossil collector turns over slabs of rock, and leaves them for the weather to wash clean.

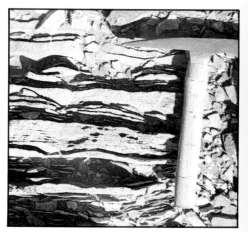

▲ A cross-section of the Ediacara sediment showing symmetrical tidal ripples.

implies that there were floating or swimming animals on which to feed; these must have been the appropriate size to be grasped by tentacles and then transferred to the stomach for digestion. There is no direct evidence of what type of animals they were.

As well as the free-swimming coelenterates there were sessile colonial forms fixed to the sea-bed. These were the sea pens, with a central axis and fronds which give the animal a bilateral symmetry. Along each subdivision of the frond were small animals, each with its own tentacles for catching small floating or swimming organisms *Xenusion*, supposedly an intermediate animal, in fact has been shown to be part of a sea pen. Some four different types of sea pens are known from Ediacara and this again implies the existence of small animals floating in the water.

Both the sea pens and jellyfish must have been passive feeders relying on their food quite literally bumping into them. Remains have also been found of various active swimming or crawling organisms such as the segmented worms, of which seven kinds are known. One form, the slow-moving *Dickinsonia*, is bilaterally symmetrical and reminiscent of the annelid worm alive today known as

the sea mouse *Aphrodite*, although it has been suggested that it was a primitive mollusc. The more clearly active form *Spriggina* has a body with up to 40 pairs of parapodia, the term given to the limbs of marine annelid worms, but the head end is clearly differentiated into a horseshoe outline reminiscent of an arthropod, which suggests it was more advanced than worms. A more recently described type of animal from Ediacara is *Praecambridium*, which is clearly an arthropod, but also has a soft covering and not the mineralized carapace normally associated with arthropods like crabs and trilobites.

As well as these body fossils, there is ample evidence of organisms that lived buried in the sediment: worm-like detritus feeders and others which grazed on the surface of the sediment. There is, however, no way of relating these trace fossils to the body fossils, although some may have been made by the same animals.

The Ediacara fauna gives us a glimpse of what kind of life existed 680 million years ago. There were floating animals and organisms which lived fixed to the substrate, there were animals that moved over the surface or swam in the water, and there were others which lived burrowing within

the sediment. Some of these animals preyed on organisms as yet unknown. Once the way of life of an animal is understood, it automatically implies the existence of other elements of the fauna, even if no actual fossil remains are preserved.

This Precambrian fauna is worldwide, occurring in south-west Africa (Namibia), Australia, England and Russia. Perhaps the most remarkable discovery was made in Charnwood Forest, Leicestershire, England, in a place often visited by geology students, who examine the Precambrian volcanic rocks and sediments. Yet in 1962 it was a schoolboy who first noticed large fossils were perfectly preserved there and visible to anyone. The professionals had not expected to find fossils because the rocks were volcanic and Precambrian in age and so did not see these fossil sea pens which were named *Charnia masoni* in honour of the schoolboy discoverer (who is now a professional geologist).

In the Precambrian era, the near-shore shallows must have been teeming with life, but little of this would have been preserved. The fossil record of the Precambrian is sparse, but as most of the animals were soft-bodied, this is perhaps to be expected.

▲ *Cyclomedusa*, a fossil jellyfish.

▲ *Spriggina*, an important fossil which can be interpreted as an intermediate stage between an annelid worm and an arthropod.

▲ *Dickinsonia*, a type of annelid or segmented worm.

▲ A fragment of *Dickinsonia*, alongside trace fossils of an unknown organism that was crawling through the fine sand.

THE BURGESS SHALE

The Burgess Shale is a unique record of a single environment frozen for a split second of geological time.

The Ediacara fauna is of special importance since it preserved a number of soft-bodied animals from the Precambrian era. Most living marine faunas cannot be expected to leave any recognizable evidence of their former existence (see pp 56–7). However, there is a unique deposit of a large variety of marine fossils from the Cambrian; this is known as the Burgess Shale.

This famous deposit was discovered quite by accident by Charles Doolittle Walcott (1850–1930). After a career of nearly 30 years as a palaeontologist of the US Geological Survey, in 1907 Walcott became Secretary of the Smithsonian Institution, Washington, and was able to devote himself to studying Cambrian rocks. In 1909 he was travelling on horseback through the Canadian Rockies in British Columbia, examining the thick Cambrian sequences there, and on a mountain trail between Mount Wapta and Mount Field, he stumbled over a fragment of shale with a perfectly preserved trilobite. The next field season Walcott

returned with his two sons and a party of helpers and they located the 1m-thick layer which contained the fossils. That year and the following summers the fossil-bearing Cambrian rock was quarried out and the slabs taken down to their camp and carefully split; between 1910 and 1917, 40–50,000 specimens were collected. Between 1911 and 1931 Walcott published a whole series of scientific papers on the collection (the last being published posthumously) in which he described some 150 species belonging to 119 genera. It was not until 1966–7 that the quarries were reopened and a new series of detailed studies began.

Although it is now thousands of metres above sea-level, 550 million years ago the Burgess Shale was deposited at the foot of a giant underwater cliff in the deep sea. The Burgess Shale is unique in the number of perfectly preserved remains of soft-bodied animals it contains. In the normal course of events, the soft parts usually rot away or are eaten by scavengers. To preserve the soft parts of animals in great detail requires very

special conditions indeed. The significance of the Burgess Shale is that it was deposited in such unusual circumstances that most animals living in the sea in that particular area were fossilized. As could be expected, there were many kinds of animals in this deposit that do not exist today and some that have no living relatives of any kind whatsoever.

How was this unique deposit caused? From the shapes of the animals as well as the types of legs or other organs they possessed, it is possible to work out their mode of life. An animal with legs can be assumed to have been free-living and capable of walking or swimming. An organism resembling a sponge is likely to have been fixed to the sea-bed. From studying the fossils, it can be shown that some animals lived on the sea-bed, some burrowed in it, others crawled over it and still others swam in the waters above. Yet in the Burgess Shale there is no sign of burrows, no sign of animal tracks and trails, nor of any organism attached to the substrate. All the animals are preserved as flattened

▲ *Opabinia*, a form unrelated to any known animal group, had five compound eyes and food catching spike bearing proboscis.

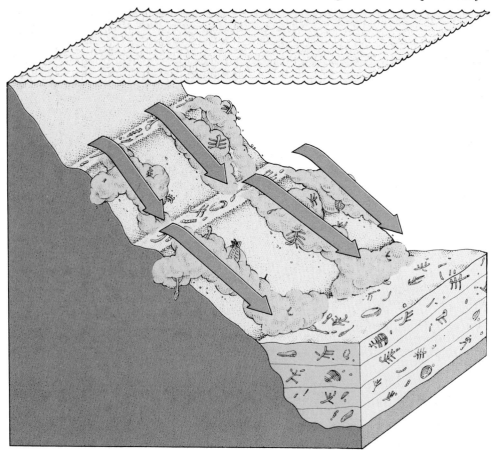

▲ The Burgess Shale is preserved at the foot of a huge underwater cliff; the animals lived on the muddy ledges. Occasionally the mud would avalanche, sweeping the fauna to the bottom and burying it.

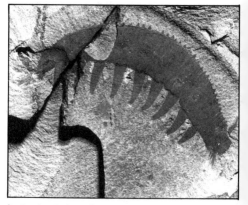

▲ *Aysheaia* has characteristics of both annelid worms and arthropods and seems to be related to the living *Peripatus*.

film due to the compaction of the fine muds. From the way in which the fossils have been squashed as the mud was compressed, it is clear that they were lying in the mud at varying angles. This means that they could not have simply sunk down onto the muddy bottom to become gradually covered by more sediment falling from above. All must have been swept down in a great muddy avalanche, so that when the mud and animals came to rest they were churned up and orientated at different angles. The mud slumps must have come to rest in deep waters where there was no oxygen and therefore no scavenging organisms to eat up the remains.

The Burgess Shale occurs in the region of a huge algal limestone reef, over 165m high, at the base of which fine muds were banked. Where the muds were banked up higher than the stagnant bottom waters, they were colonized by animal life. Fixed on the surface of the mud lived a number of sessile organisms – some 15 different genera of sponges as well as coral-like animals. One of the sponges, *Pirania*, had long sharp siliceous spikes to which primitive lamp shells, or brachiopods, attached themselves. In this way the lamp shells were protected and at the same time could filter their food from the clearer waters well away from the muddy bottom

sediment. There were primitive crinoids, or sea-lilies, and an animal called *Dinomischus* which belongs to a completely unknown division of the animal kingdom.

Living within the sediment, either tunnelling through it or simply hidden in burrows, were a great variety of worms. Among the more important of the worms at that time were the priapulids, which are very rare today. There are only two living genera but in the Burgess Shale there were at least seven. These worms have a hook-bearing proboscis, which can be protruded to capture prey. The victim is then passed into a muscular pharynx containing chitinous teeth which break it down. One type of priapulid, *Ottoia*, abundant in the Burgess Shale, fed on small shellfish, as well as worms – including, occasionally, even its own kind. The diet of these priapulid worms is known with certainty because their gut contents are preserved.

Swimming over the muddy surface were polychaete annelid worms with fine tufts of bristles which helped them to swim. *Canadia* is one of the more exquisitely preserved of these forms. The minute details of the delicate appendages are preserved because, when the animal was swept along in a slurry of mud, fine particles penetrated in amongst all the fine hair-like projections. Crawling over the

surface of the muds were a number of primitive molluscs, some looking like the Chinaman Hat limpets and some such as *Wiwaxia* with scales and protective spines.

By far the most abundant animals of all were the joint-limbed arthropods, which must have spent most of their lives crawling over the sea-bed, feeding on small particles of organic matter in the mud. The most familiar arthropods of this period were the trilobites and their relatives. Trilobites are common fossils as their carapaces, comprising a head shield with compound eyes, jointed thorax and a tail, are easily preserved. Detailed information on their appendages is much rarer. On each body segment there was a pair of jointed walking legs as well as a pair of gill filaments for respiration. The main part of the limb closest to the body bore sharp spines, which could shred small animals such as worms and then pass the pieces up to the mouth.

Among other superbly preserved arthropods are *Waptia* which is similar to a fairy shrimp, and *Emeraldella* which is like a trilobite with a long tail spine. *Canadaspis* was a primitive crustacean. One of the most attractive swimmers was the tiny *Marrella*, with enormous spines over its back, long walking legs and delicate feathery gills. Although many of these animals

▲ *Marrella* is a small spiny primitive arthropod.

▲ *Canadia*, a polychaete worm.

▲ *Hallucigenia*, a bizarre creature of unknown affinities with seven pairs of legs – possibly a scavenger.

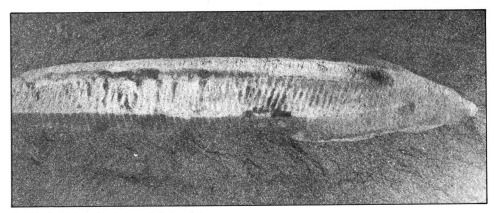

▲ *Pikaia*, the earliest recorded chordate, which used its V-shaped array of muscles for swimming.

are difficult to classify, there is no doubt that they are primitive members of the arthropod phylum.

Classification becomes more difficult with one of the most famous animals in the Burgess Shale, *Aysheaia*, which is a kind of missing link between arthropods and annelid worms, in the group the onychophores. There is a living member of this group, *Peripatus*, with velvety skin and stubby limbs bearing small claws, which lives under the bark of trees in tropical forests. *Aysheaia* seems to have been a marine form of *Peripatus*, with limbs that carried sharp spikes. The first pair of appendages had six spikes, the walking legs had a pair of forward-pointing spikes, but the last two pairs had a single backward-pointing spike. These fossils are associated with sponge remains and it seems likely that with their claws the animals clambered over sponges, anchoring themselves with their spikes and using the first pair of appendages for tearing into the sponge on which they fed.

In the Burgess Shale there are some animals that are so weird that if it were not for the conclusive evidence of many perfectly preserved specimens, no one would believe such things have existed. Perhaps the most bizarre is the aptly named *Hallucigenia*, which seems to have moved along on the tips of seven pairs of stiff spines. Along the mid-line of its cylindrical body were seven flexible tentacles, the tips of which were divided – their function remains a mystery. These animals are thought to have been scavengers; 15 individuals have been found associated with a large worm, which suggests that they may have gathered to feast on its corpse.

Yet another strange creature is *Opabinia*, a segmented swimmer with flat overlapping plates carrying gill filaments running along either side of the body. The plates seem to have been used for swimming over the surface of the sea-bed. The last three segments are turned upwards, perhaps to serve as a rudder. On the top of the head were five compound eyes on short thick stalks. Protruding from the front of the head was a large flexible trunk with grasping spikes at its tip. This was for capturing prey, which was then conveyed to the mouth situated on the underside of the first segment.

All the animals preserved in the Burgess Shale were accidentally caught up in mud that slid off parts of the underwater cliff face and were swept down into the lifeless depths. Animals swimming close to the bottom, as well as those living in the mud, stood a very good chance of getting

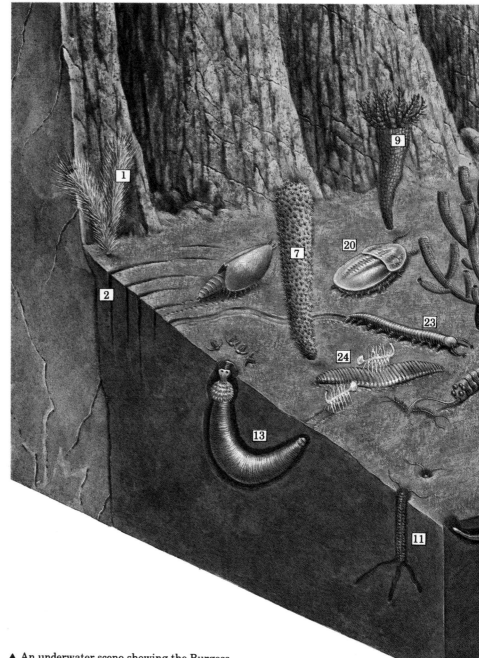

▲ An underwater scene showing the Burgess Shale fauna as it might have appeared in life.

▲ The Burgess Shale, situated high in the Canadian Rockies.

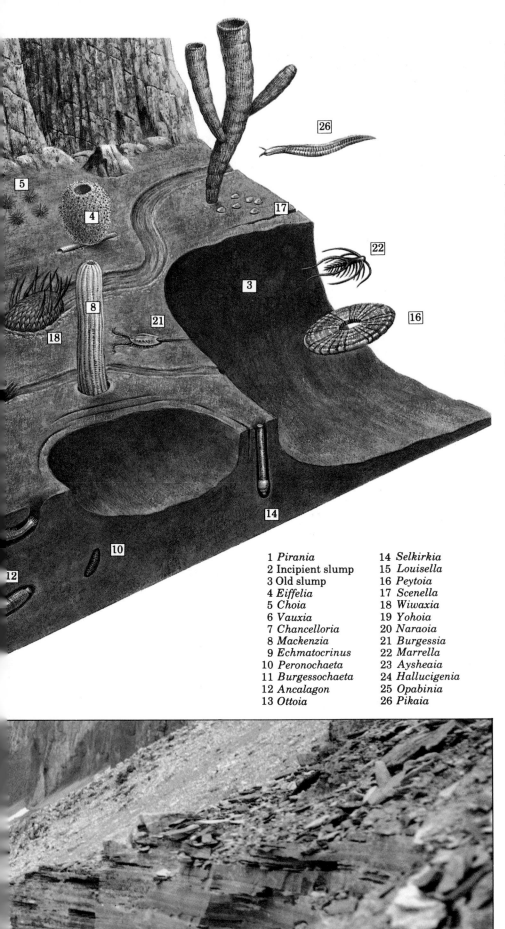

1 *Pirania*
2 Incipient slump
3 Old slump
4 *Eiffelia*
5 *Choia*
6 *Vauxia*
7 *Chancelloria*
8 *Mackenzia*
9 *Echmatocrinus*
10 *Peronochaeta*
11 *Burgessochaeta*
12 *Ancalagon*
13 *Ottoia*
14 *Selkirkia*
15 *Louisella*
16 *Peytoia*
17 *Scenella*
18 *Wiwaxia*
19 *Yohoia*
20 *Naraoia*
21 *Burgessia*
22 *Marrella*
23 *Aysheaia*
24 *Hallucigenia*
25 *Opabinia*
26 *Pikaia*

buried in this way. On the other hand, animals that lived in the upper layers of the sea would rarely have been caught up in such muddy avalanches. Nevertheless, a few of such animals were unlucky enough to get trapped. They include the hunting arrow-worm *Amiskwia*, with its lateral and tail fins. One of the most important fossils is *Pikaia*, known from 30 specimens, which appears to be a distant relative of the vertebrates. It seems to have had a notochord (which became the backbone in the later vertebrates) and its muscles were arranged in a V-pattern characteristic of fishes. This small fish-like animal is similar to the living lancelet, or *Amphioxus*, which inhabits the sandy sea-bed in many parts of the world. *Pikaia* may well be the first representative of the group which gave rise to all the backboned animals, including man.

The most striking aspect of the organisms that inhabited the bottom sediment is that they were dominated by priapulid worms which today are restricted to cold waters and often inhabit anaerobic muds. In the modern seas their role as scavengers is taken by polychaete annelid worms, which in Ordovician times (500 million years ago) evolved chitinous jaws. Today they burrow, crawl and swim, and live in vertical burrows in a wide range of marine habitats, having completely ousted the priapulids.

The vagrant or wandering fauna was dominated by arthropods, which made up 43% of species in the Burgess Shale, and it is important to realize that many of the curious forms preserved have extremely fragile cuticles, or exoskeletons, that normally would not be preserved. The inhabitants within the bottom sediment were dominated by priapulids (12% of species). The change from the Burgess Shale life pattern to modern marine conditions cannot be traced because there is no other well-preserved example of a cross-section of marine life in the fossil record. One can only emphasize that the situation has changed enormously.

The preservation of these soft-bodied animals has given us a single glimpse of the vast range of fauna normally missing from the fossil record. This unique deposit also reminds us that an ancient environment with living things is dependent upon a number of creatures and plants that are not represented directly in the rocks. Other rare rock formations exist which add new dimensions to our understanding of past life, but none to such a dramatic extent as the Burgess Shale.

THE INVENTION OF SKELETONS

Skeletons are quite easily preserved and are responsible for the abundance of fossils in the Cambrian. But how did skeletons arise in the first place?

The abundance of the fossil record at the beginning of the Cambrian era gives a strong impression of a sudden explosion of animal life. However, it did not necessarily happen that way. The reason for this sudden increase in fossils is that many different and quite unrelated forms of life began to develop mineralized skeletons, and thus were more likely to be preserved than the soft-bodied animals which preceded them. The question that arises is how, in such a comparatively short period of time, so many different groups of animals were able to develop skeletons, both externally and internally.

One line of approach is to consider how skeletons are formed today, as this may provide a clue as to how they arose in the first instance. The relevant information is to be found in disciplines seemingly unrelated to palaeontology, such as biochemistry and the medical sciences. This theory of the origin of skeletons is entirely speculative, although there is circumstantial evidence in its support. It arose out of an apparent paradox.

Bone exists in a dynamic equilibrium with blood. This subject is of major concern to the medical profession, and the regulation of the correct amount of calcium in the blood is critical for the proper functioning of, for example, the nervous system. Bone is able to act as an important chemical reservoir. Bone mineral is calcium phosphate; if it is shaken in powdered form in a salt solution of the same concentration as blood plasma (or serum) it will dissolve. In fact, at half the concentration found in blood serum, calcium phosphate will dissolve in a solution. On the other hand, it takes over double the plasma concentration of calcium phosphate before bone mineral comes out of solution. Two questions arise: why does the skeleton not dissolve away when in contact with the serum, and how can bone mineral be extracted from the low concentrations at blood serum levels?

The answer to these questions came in 1955 from the classic experiments of W. F. Neuman and M. W. Neuman. They added collagen, the basic structural protein of bones and teeth, to a solution of calcium phosphate at the same concentration as in plasma and immediately bone mineral was formed. It appeared that collagen formed the template on which bone mineral crystallized, with the individual molecules dropping into place on the crystal lattice. This was an alternative process to precipitation, which involved concentrating the mixture so that 18 molecules had to come together simultaneously. This explained the paradox regarding blood plasma concentrations.

The remarkable ease with which calcium phosphate mineral was formed in the presence of collagen posed an even greater problem. How did collagen manage to avoid becoming mineralized, bearing in mind that a third of the body's protein is collagen? The next part of the experiment involved adding a substance from the blood plasma known as 'ultrafiltrate' to a calcium phosphate solution, then adding collagen; the ultrafiltrate prevented the mineral from being formed. It was then found that there were organic phosphates in blood

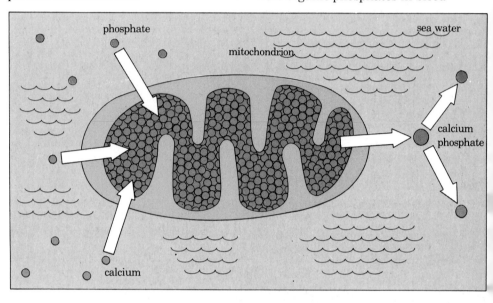

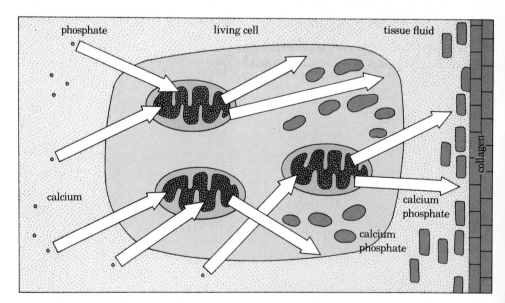

▲ **The role of the mitochondrion in mineralization.** The upper diagram shows a mitochondrion as an independent living organism in the sea. Calcium and phosphate ions enter the mitochondrion, and combine to form the mineral calcium phosphate which is then ejected and subsequently breaks down into separate calcium and phosphate ions. The lower diagram shows calcium and phosphate entering a living cell to be absorbed by the mitochondria. The resulting calcium phosphate is then ejected and may remain in the cell, or may be pumped out to be deposited on an organic template, such as collagen, to form bone.

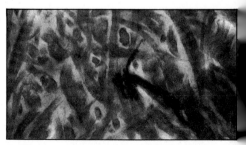

▲ Electron micrograph (× 12,500), showing calcium phosphate crystals growing in assocation with collagen.

plasma which acted as crystal poisons. These molecules sat on the collagen and blocked the seeding sites of bone mineral. Where mineralization took place, there was an enzyme present that removed the crystal poisons and hence allowed the near automatic deposition of mineral to proceed on the collagen template.

It became generally accepted that the clue to mineralization lay in the special properties of the protein collagen. But in the early 1960s it was discovered that, in cartilage, the crystallites of the bone mineral calcium phosphate were not seeded onto collagen, but onto a completely different kind of molecule, not even a protein. It transpired that other molecules such as keratin, the protein of hair and nails, were also capable of seeding minerals. This meant that there must be a more widespread phenomenon of mineralization that was not dependent on the structure of any particular protein or complex organic molecule.

One of the first clues came from a physiologist researching eggshell development; he found that all too frequently his microscopic preparations would be ruined because certain organelles within the cells, in particular the mitochondria, would suddenly soak up calcium. In Lehninger's textbook of biochemistry, it was noted in passing that, in the liver, the mitochondria were loaded with mineral granules. The mitochondria, which are present in every body cell, seemed to have an affinity for calcium and phosphate ions in much the same proportions as those in bone mineral. Two research workers at the Royal Dental Hospital, London – Irving Shapiro and John Greenspan – in 1969 were the first to outline the theory of the key role of the mitochondrion in mineralization, noting that the hormones that affected bone mineral in plasma levels and bone resorption and deposition acted primarily on the mitochondrion. This was the clue; it looked as if the mitochondrion, the powerhouse of the cell, was responsible for mineralization.

The next question was why the mitochondrion should be concerned with collecting up and then expelling calcium phosphate. There was a possible evolutionary explanation. One of the substances in sea water that automatically enters living things is calcium and there needs to be some active pumping mechanism to get rid of it. This waste disposal function of the mitochondria is likely to have been an important and necessary role long before they became incorporated into the eukaryote cell. It begins to look as if the very initiation of skeletal formation was a means of dealing with excess calcium; one of the easiest ways of doing this was combining it with phosphate and expelling it. Later in evolution other groups of animals developed skeletons of calcium carbonate, but usually the first kind of mineralization was phosphatic. Calcium carbonate skeleton formation is more complex.

This theory accounting for the origin of skeletons still leaves unanswered why it should have occurred over a period of a few million years about 570 million years ago. It might have been related to the gradual decline in limestone-forming stromatolites. This implies that the amount of calcium in the oceans was substantially increased, as it was no longer being removed to form limestones, and at a critical juncture one of the most effective means of getting rid of it may have been by depositing calcium salts in the form of skeletons. Once this had been accomplished, the geological record had an abundance of fossils.

▲ A nest of *Protoceratops* eggs from the Gobi Desert. Eggshell, the external skeleton of an egg, is composed of calcium carbonate.

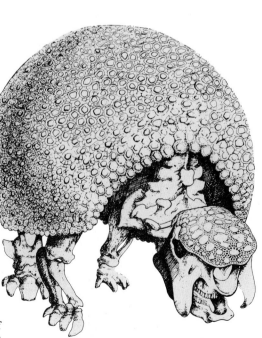

▲ The Pleistocene *Glyptodon*, an extinct relative of the armadillos. It was a member of the only mammalian group to possess both an internal and external bony skeleton.

THE HISTORY OF SHELLFISH

Shellfish have the best recorded history of all animals. Some of the most primitive forms are still alive today.

Marine sedimentary rocks are usually abundant with fossils. They bear witness to the types of animal that have lived under the sea since the Cambrian, 570 million years ago. The number of environments under the sea are limited and so are the feeding strategies of the animals which exploit them. The fossil record of the past 570 million years demonstrates the successive replacement of one type of shellfish in a particular food niche by better adapted forms, so that the major eras are characterized by the dominance of different animal groups.

One of the most significant events at the beginning of the Cambrian was the emergence of organisms from their burrowing life in the sediment, perhaps as a consequence of their acquisition of protective skeletons. It is assumed that worms of some kind inhabited the vertical burrows that are common trace fossils. These worms probably extended their food-gathering organs and filtered nutrients from the sea water. But the major advance came when the food-filtering devices were raised clear of the bottom sediment (although the Precambrian sea pens had previously accomplished this).

Organisms raised clear of the substrate easily succumb to the attentions of any predator, unless protected in some way such as by a shell. Yet there were extremely vulnerable Cambrian sponges and sponge-like colonial organisms, the archaeocyathids, that formed large inverted cone-like structures fixed on the sea-bed, which seem to have been highly successful. Another advance was found among the coelenterates. These were the corals which developed the ability to put down layers of calcium carbonate so that the food-collecting tentacles could be raised higher and higher in order to reach the necessary nutrients more easily.

The brachiopods or lamp shells also extracted food by filtering minute organisms from sea water. They possessed long, coiled, hairy, food-gathering organs, called 'lophophores', and were fixed to the substrate by means of a fleshy stalk or pedicle. They possessed two hinged valves, the more primitive with a calcium phosphate shell, the more advanced with calcium carbonate. The most primitive type of brachiopod from the Cambrian is still living. This *Lingula* is very adaptable and can live in brackish water conditions. It inhabits vertical burrows in the sediment and is found around the coasts of most continents although it is most numerous in the Far East. All the more advanced brachiopods lived with their shells out of the sediment.

Another important group of animals in the Cambrian that fed on microscopic organisms were the echinoderms or spiny-skinned animals like sea-urchins, with a skeleton just under the skin of calcium carbonate plates. The primitive examples seem to have been little more than hollow balls with a mouth and anus. The surface of some became covered in food-gathering grooves, whereas others developed small 'arms'. The plant-like creatures, the crinoids or sea-lilies, grew stalks which lifted their arms still further off

▲ Silurian nautiloid shells from Czechoslovakia: the simple straight shell of

Ormoceras pelucidum shows longitudinal banding; the two curved shells are of

▶ A reconstruction of the Silurian sea-bed fauna based on the Budňany Beds, near Prague, Czechoslovakia. The 'zigzag' and 'striped' patterning of the cephalopods is actually preserved in the fossils. Also shown are trilobites, brachiopods, crinoids (sea-lilies) and corals.

the substrate. Numerous ancient limestones are composed almost entirely of the calcareous skeletons of these crinoids.

Fixed sessile organisms must have had a free-swimming or floating larval stage, for without this the distribution of such animals over vast regions would be incomprehensible. The larvae of sessile animals constituted a considerable proportion of the plankton – the organisms inhabiting the upper regions of the seas. In this way they

provided much of the food for sea creatures which filtered the water or had tentacles for catching their food. Larger free-swimming animals such as worms and primitive chordates, fish-like animals similar to the living lancelet, also fed on the plankton.

Aside from sessile and swimming creatures, there were those which wandered along the sea-floor in search of food. The main source of their food was organic material accumulating on the surface of the sea-bed – detritus.

Among the most successful of detritus-feeders were the arthropods, or joint-legged animals, which have an external skeleton, the best known of which are the trilobites. The early types ploughed through the surface layers of the sediment leaving characteristic feeding furrows (see pp 50–1). The possession of a rigid external skeleton meant that the trilobite had to shed its skin periodically in order to grow and this made it an ideal fossil producer. There were lines of weakness in the head region along which the carapace split open and the animal moulted. During their evolution they became adapted to different modes of life: some became blind and very smooth and were clearly burrowers; some developed remarkably advanced eyes so clearly were not burrowers; others were blind but had enormous spines, a median one pointing forwards and two lateral ones pointing backwards; and there were also forms covered in intricate delicate spines. The spines may simply have been for protection, but it is not possible to be certain about the meaning of all the specializations. However, it is clear from the fossil record that the trilobites were very successful in the Lower Cambrian.

The second major group of mobile animals were the molluscs. These animals comprise a large muscular foot with a head and a back covered by a protective shell. The most primitive types of mollusc, the monoplacophorans, were thought to have died out in the Ordovician, about 500 million years ago, as there is no record of them after this. But live specimens of one, *Neopilina*, were dredged up out of the Pacific Ocean in the early 1950s. This was an exciting discovery as the shell showed a segmental pattern unknown in molluscs, but found in annelids and arthropods, suggesting that *Neopilina* was a link.

A primitive mollusc moved over the substrate by waves of muscular contractions of the foot and fed on algae and other material which it was able to rasp off the surface by means of a tongue with rows of minute teeth. From this basic type of mollusc came three major lines of development which were clearly separated by the time of the Ordovician. The line that retained the same basic feeding strategy was the gastropods or snails which are notable for having undergone a twisting of the body so that the alimentary canal is turned round and the cavity in which it opens and in which the gills are housed faces forwards into the current. This makes

Rizosceras parvulum and the coiled shell is *Peismoceras pulchrum*.

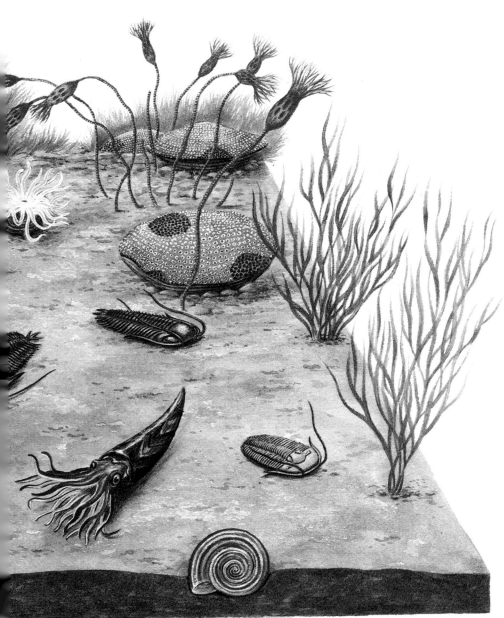

it easier for water to pass over the gills for respiration than if they were placed behind. In a second line of development the shell became double, hence the name bivalve. Both the foot and head were reduced and the gills became a food-filtering device. The third evolutionary lineage, well-established by the Silurian, led to the most active and most advanced of all the animals without backbones, the cephalopods, such as squids and octopuses. In these the foot was divided into long muscular tentacles. But the most important feature of all is that these animals were able to lift off from the bottom and swim because their conical shells accumulated gas for buoyancy.

The first cephalopods, the nautiloids, developed long conical shells with partitions separating the gas chambers. These animals drew water into the gill-containing mantle cavity, and expelled it via a tubular muscular outlet or siphon; the outflow could be directed so that the animal could drive itself through the water by means of jet propulsion. During the Palaeozoic era (570–225 million years ago) these animals were the major predators of the oceans, feeding on arthropods and other free-living animals.

There was, however, one serious problem which faced the early nautiloids: they needed to be able to move through the water, head forwards and body behind, rather than floating with the shell pointing upwards and the head and tentacles dangling down. This problem was overcome by weighting the shell by secreting and depositing calcium carbonate within it.

The early nautiloids are classified by the different patterns of shell thickening that they produced. The effect of the thickenings was to ensure that in conjunction with the gas-filled chambers or camerae, the shell was held horizontally. The cone became coiled round simply to ensure that the buoyancy gas-filled portion was above the body chamber and that as the animal grew and new chambers were added on, the orientation of the animal would remain the same.

The other feature that developed was the shape of the calcareous partitions between the chambers. In the early forms this wall or septum was a simple smooth concave shape, but as time went on these septa became more and more bumpy and wrinkled. This helped the body fit more

▲ The ammonite *Echioceras varicostatum*. The shell is broken in many places and the complicated sutures which represent the divisions between the chambers can be seen.

firmly onto the septal wall. Where the outer shell met the edge of the septa complicated folded patterns were formed called 'sutures', because they appear superficially like the sutures between bones of the skull.

The advanced cephalopods with these sutures were known as ammonoids, but at the end of the Permian period, 225 million years ago, all but one family of ammonoids died out. With the beginning of the succeeding Mesozoic era, from the Triassic through to the end of the Cretaceous (225–64 million years ago), this one surviving ammonoid family underwent an enormous evolutionary radiation to produce the huge number and variety of ammonoids which dominated the Mesozoic seas. Ammonoids were free-swimming ocean-going animals and so had a very wide geographical distribution. They seemed to undergo very rapid changes in the shape and ornamentation of their shells, so that separate evolutionary lineages can be traced in great detail. Perhaps the most famous example is that of the fossil ammonite *Kosmoceras* described by R. Brinkmann in 1929, who collected all the specimens layer by layer from a clay-pit near Peterborough, England. The changes in size and ornamentation of the shell demonstrated its slow and gradual evolution.

Another group of cephalopods in the Mesozoic were the belemnites, close relatives of the cuttlefish known from the fossils of their internal skeletons. These are bullet-shaped structures of solid calcite called 'guards'. There was a thin shell extending from the front end of the guard, called the 'pro-ostracum'. Fossils were found in Lyme Regis during the early nineteenth century in which this pro-ostracum is preserved with ink-sacs and also with tentacles which are armed with rows of small chitinous hooks. It has always been assumed that these fossils represented the soft parts of belemnites. It is known that giant marine reptiles such as ichthyosaurs fed on such animals because there are vast accumulations of the chitinous hooks preserved in their stomachs. The guardless soft body may have been preserved by the ichthyosaurs spitting it out by mistake, after which it sank into deeper stagnant waters and was preserved in the anoxic muds.

It had always been expected that

▲ *Australocerus gigas*, un uncoiled ammonite.

▲ Drawings of fossil belemnite ink sacs from *The Bridgewater Treatise* (1836) by William Buckland. The fossil ink was used to make the drawings.

▲ Ammonites preserved in the Lower Jurassic rocks of Charmouth, Dorset, England. The ammonites are at different stages of growth and were preserved in the sea-bed of nearly 200 million years ago.

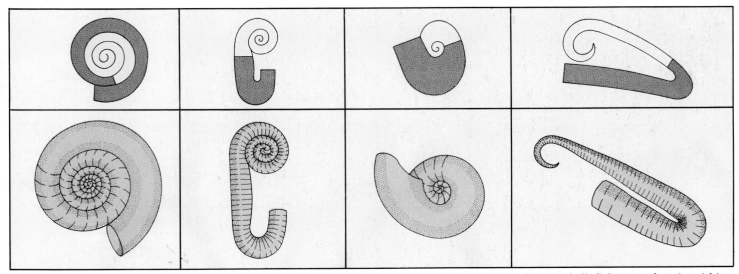

▲ **Different patterns of ammonite coiling** Ammonites usually had a single coil, but in the later forms there was a wide variety of coiling, as shown in the lower part of the diagram. The shaded portion in the upper drawings shows the position of each animal's body in its shell. Calcareous deposits within the shell weighted it so that it was in the correct position.

eventually belemnite fossils with both guard and soft parts would be discovered. And so they were in the 1970s in a privately owned quarry in southern Germany. These fossils found their way into several museums, having been purchased for large sums of money. They seemed to confirm what had been anticipated, until a student noticed that the calcite guard appeared to be carefully glued onto rocks showing the preserved soft parts. In fact the association of the guard and soft parts was a deliberate fraud, providing the palaeontologists with what they wanted. Nothing had been confirmed after all. In fact, a complete fossil belemnite has still to be found.

The trilobites became extinct during the Permian crisis and during the succeeding Mesozoic other kinds of arthropods, especially the crustaceans, such as lobsters and crabs, became the main scavengers of the sea-bed; they were not merely detritus-feeders but consumers of carrion of all kinds. The crustaceans developed the ability to construct complicated living galleries within the sediment (see pp 52–3). These acted as refuges from enemies and also allowed the crustaceans to extract nutrients from the sediment while remaining hidden from view. Among the fixed filter-feeders, the sessile crinoids continued; however, the heart-urchins, which had inhabited burrows in which they set up water currents to bring in food particles,

evolved into free-living sea-urchins.

The other filter-feeding shellfish, the brachiopods, formed a much less prominent part of marine life in the Mesozoic. There were two main kinds, the ribbed rhynchonellids and the smooth terebratulids, the typical lamp shell resembling an ancient Roman lamp. The ecological niche of the brachiopods was being taken over by the bivalves, which had the advantage of being able to burrow in the sediment and hide from predators.

The Late Cretaceous extinction, which sealed the fate of the dinosaurs (see pp 146–7), saw a drastic reduction in the ammonite faunas as well, and their geographical distribution became more and more restricted. Towards the end there was only a handful of forms restricted to small isolated tropical seas. In the succeeding Cenozoic or Tertiary era the ammonite niche was occupied by the direct descendants of the belemnites. Intermediate forms are known which link the belemnites with the main groups of living ocean-going cephalopods, such as *Loligo* the squid and *Sepia* the cuttlefish. Only the ancestry of the *Octopus* cannot be traced because it has no internal skeletal parts. In the squid only the 'pen' or pro-ostracum remains, while in the cuttlefish the guard is represented by the cuttlebone which is made up of numerous gas chambers, which give it a light texture(cage-birds and seagulls peck it in order to obtain

▲ Internal and external view of a bivalve shell with a neat circular hole, which has been bored by a carnivorous gastropod (snail).

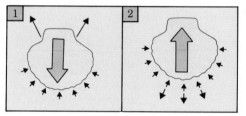

▲ 1 The normal swimming action of a scallop. 2 The scallop's escape reaction causes it to move in a different direction.

▼ **Some different bivalve ways of life**. 1 *Mya* (gaper clam) is buried deep in the sediment with fat permanent siphons. 2 *Solen* (razor shell) is buried in the sediment. 3 *Scrobicularia* has long siphons that can be withdrawn into the shell. 4 *Pinna* (pen shell) is fixed in the sediment by long byssal threads. 5 *Mytilus* (mussel) is fixed onto the rocks by byssal threads. 6 *Pecten* (scallop) swims by drawing

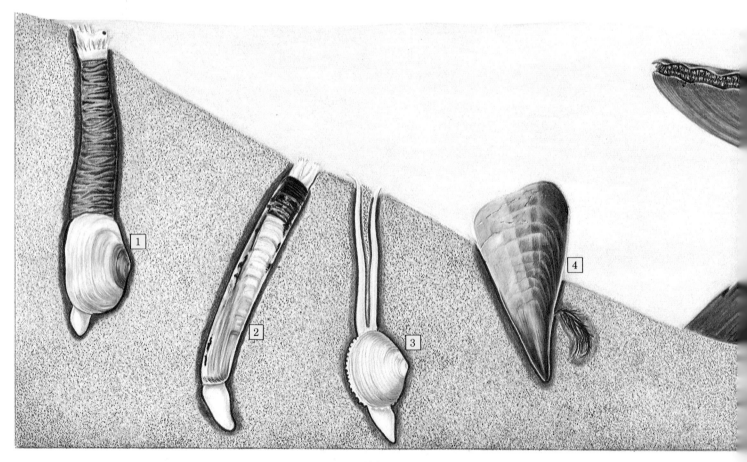

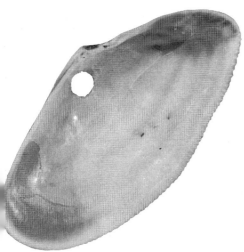

This is evidence of the development of predatory habits by gastropods, which initially were herbivorous.

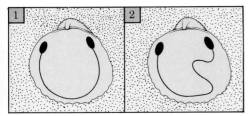

▲ 1 Cockles with short siphons live just beneath the surface. 2 Those with long siphons live deep in the sediment.

water into its cavity and then expelling it. 7 *Teredo* (ship-worm) bores into wood by means of its shell and lines the burrow with a calcareous deposit. 8 *Ostrea* (oyster) is cemented onto the rock by the left valve. 9 *Pholas* (piddock) bores into rocks. 10 *Gryphaea*, an extinct oyster, that sat on the soft substrate. 11 *Cardium* (cockle) burrows just beneath the surface of the sediment.

calcium salts). As active, aggressive predators, cephalopods remain among the most successful of all the animals without backbones and mark the peak of the evolution of invertebrates.

However, the main bulk of Cenozoic fossils are gastropods and bivalves. The bivalves had taken over from the brachiopods, which survive today in their advanced form only in the coastal waters around New Zealand because of its geographical isolation from the rest of the world.

The structure of a bivalve's shell reflects accurately the conditions in which it lived and so bivalves are especially important to palaeontologists as indicators of different environments. Primitive bivalves, such as the living *Nucula*, crawl in a snail-like fashion through the sediment, ploughing a shallow groove as they go. Water is drawn into the mantle cavity where the gills are situated; the gills are used to collect particles of food as well as for respiration. In more advanced bivalves, two small tube-like orifices are formed which connect the mantle cavity to sea water. Water with food particles is drawn in through one tube and expelled through the other. In those bivalves that live buried deep in the sediment, the mantle is drawn out to form long tubes or siphons which work in the same way. In order that these can be safely withdrawn into the shell, the attachment of the mantle to the shell has a large indentation or pocket

for the accommodation of long siphons, known as the 'pallial sinus'. When these are present in a fossil they can be taken as evidence of retractable siphons and, therefore, a bivalve which lived buried in the sediment.

Some bivalves protect themselves by boring into rock or into wood; in the latter case *Teredo*, the ship-worm, uses its two valves rather like a bit to cut its way into the wood. Other protective strategies include living fixed to rocks in the intertidal zone in colonies, as do mussels, fixed by horny threads (the 'beard') secreted by a reduced foot. Some, such as oysters, develop exceedingly thick shells and cement themselves onto rocks. There are even swimming bivalves, the scallops, which move in a bouncing saucer-like fashion by drawing water in through the wider part of the shell away from the hinge and expelling it in two narrow streams on either side of the hinge. This is how they swim normally, but when danger threatens, they open their shells and then clap them together, so that water is forced out suddenly and they shoot off at speed.

Much of the fossil record of the past 64 million years consists of shellfish that have living representatives or very near relatives in our seas today, so it is not difficult to reconstruct their habitats and way of life. Following this uniformitarian approach will bring us close to the right interpretations.

FRESHWATERS

Life originated in the oceans and millions of years passed before the land was colonized. But in the Devonian period evidence of the first freshwater life can be found in the Old Red Sandstone.

Life first developed in the seas, and the greater part of the fossil record comprises marine organisms. Later, of course, life spread to other environments, but initially, for all living things, the edge of the sea marked the limit of the habitable universe. Today numerous marine animals are capable of living in the zone between tides and a very small number can survive in the narrow 'splash' zone which only receives salt-water splashes from the waves at high water. To marine animals the shore has always presented a fundamental barrier to advancing onto the land.

There was, however, another type of aquatic environment other than the sea available to living things and that

was freshwater. As soon as the formation of the oceans and atmosphere had occurred, freshwater fell to earth as rain and formed streams and rivers. Alluvial river deposits are known well into Precambrian times. Yet life probably did not invade freshwaters until 400 million years ago, for it was not until then that fossils are found in freshwater sediments.

If sediments are preserved within a river system, as in an ancient alluvial plain, they must be of freshwater origin. The first question is how are fluviatile or river deposits recognized? Taking any individual piece of rock such as a sandstone, it is not always easy to work out its

environment of deposition. This can only be done if the rock is looked at in the context of the strata.

The first task is to study the conditions in modern-day rivers. These are best observed in rivers in regions that have a dry season so that the river deposits can be easily examined. A river meandering through a flood plain deposits sediment so that layers are built up in a particular sequence (see pp 22–3). Although gravels, sand and muds occur in many environments, strata comprising a gravel followed by a cross-bedded sand and capped by silts or muds that often have sun-cracks are the hallmark of an alluvial or fluviatile regime. Whenever such a sequence is preserved it is

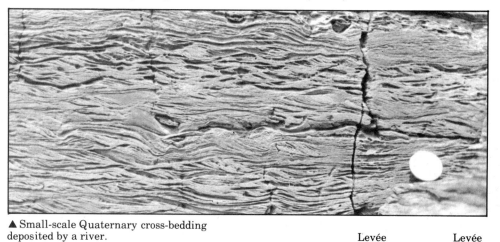

▲ Small-scale Quaternary cross-bedding deposited by a river.

▲ Large-scale cross-bedding in the Old Red Sandstone, deposited by an ancient river.

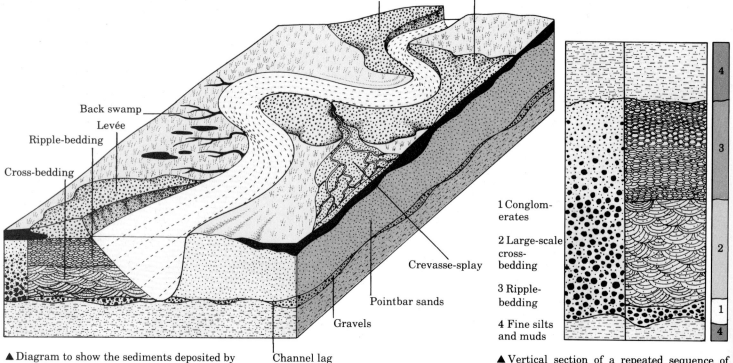

▲ Diagram to show the sediments deposited by a meandering river.

1 Conglomerates

2 Large-scale cross-bedding

3 Ripple-bedding

4 Fine silts and muds

▲ Vertical section of a repeated sequence of alluvial sediments.

evidence of a freshwater river system.

The direct comparison of modern-day flood plain sediments and those in ancient rocks leads to the conclusion that they must have been formed in exactly the same kind of environments. This approach puts into practice Charles Lyell's dictum, 'the present is the key to the past'.

By far the most intensively studied fluviatile rocks are those of the Old Red Sandstone of the Welsh Borderland (see pp 114–17). Until detailed studies were begun 25 years ago, there had been much speculation about the environment represented by these sediments. The research revealed that the rocks were about 400 million years old and contained the remains of fossil plants, arthropods and early vertebrates which provided the first evidence of the invasion of freshwaters. The Old Red Sandstone contains the record of freshwater life of the Devonian.

The sudden appearance of fossil fishes in these 400-million-year-old rocks raises the question of whether freshwater life is preserved in more ancient rocks. But this is not the case. Precambrian, Cambrian, Ordovician and Silurian freshwater alluvial sediments exist and they are all notable for the absence of fossil remains.

Comparatively little attention has been paid to the Old Red Sandstone in the past because the rocks are notoriously lacking in fossils. Yet very close studies of such apparently unpromising rocks can be enormously rewarding. There are thousands of metres of red sandstones with coarse gravel beds and red siltstones – the same sequences were repeated interminably. By measuring the slope of the cross-bedding, the direction of the flow of the water can be determined. An examination of the rock types that make up the pebbles in the gravel can reveal the geology of the land that was being eroded and can often determine the source of the river system. Remains of fossils occur as fragments of carapace or bony armour mixed up with the pebbles of the bottom deposits, or sand-grain-sized scales which behave simply as sand grains. The latter can be extracted from the rocks using dilute acetic acid (20%); the acid removes the cement of calcium carbonate so that the sand grains and fossils become separated. Under a binocular microscope the fossil scales can be picked out, using a fine paint-brush.

Plant material often accumulated in the fine-grained deposits which formed as the rivers overflowed their banks. It is in these conditions too that complete remains of arthropods and fishes are best preserved. They were swept downstream in turbid flood waters which then spread over the land as the floods burst the river banks. When the flood waters receded, the remains of fishes were preserved intact in the fine-grained sediments. The fact that animals or plants were swept downstream and were still preserved within the fluviatile regime is conclusive evidence that they must have been freshwater-living.

▲ Aerial view of the meandering Notikewin river, Alberta, Canada, showing succession of point bar sands.

THE FIRST FRESHWATER LIFE

Fish originated in the sea but adapted to freshwater. Although no evidence of freshwater plants exists, they must have been present in the environment as food for the fish.

The fossil record of freshwater organisms is extensive as far as molluscs, arthropods and fishes are concerned. Their existence implies the presence of freshwater plants as the base of the food chain in every environment is always plant life (see pp 124–7). Although there is no direct fossil evidence of the very first freshwater plants, it is assumed that such plants evolved from marine algae, but apart from those algae that deposit calcium carbonate, they are virtually unknown as fossils.

The fossil plants that occur with the arthropods and fishes represent a further evolutionary stage: they are the first land plants, which have already developed woody tissue and resistant spores which are easily fossilized. It was at the edges of rivers and streams, or any body of freshwater, that these plants were to be found. Plants need minerals and water which are present in the freshwater environment, but they also require carbon dioxide which is found in abundance in the atmosphere. The great evolutionary development of the plants was to obtain one set of requirements from their watery environment and the other from the atmosphere.

The best-known examples of early plants come from Scotland, near Aberdeen, in a rock formation known as the Rhynie Chert. This deposit was discovered by the geologist W. Mackie when he was making a geological map of the region before the First World War and a detailed study was undertaken by two palaeontologists between 1917 and 1921. The deposit is unique in that it represents what must have been a peat bog, which was suddenly inundated by hot siliceous waters, from nearby volcanic eruptions, which took place about 400 million

years ago during the great Caledonian mountain-building period. All the animals, fungi and plants were killed, but before decay could take place, a silica gel formed and solidified so that every microscopic detail of the animals and plants was preserved. The plants had water-conducting woody cells and so they are known as 'vascular plants'. On the stems and leaves there were special gas exchange pores or 'stomata' that opened and closed, which means

▼ A reconstruction of a section of the Rhynie peat bog, Aberdeenshire, Scotland. 400 million years ago this site was in a volcanic area and the plants and animals were overwhelmed by hot siliceous waters. The fossils are exceptionally well preserved and give a unique picture of Lower Devonian plants and animals.

Lepidocaris

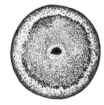

Rhynia

they exchanged gases in the air and not under water. Unlike algae, these plants had a waxy waterproofing or cuticle which prevented them losing too much water to the air. Finally, there were minute spores that were evidently windborne. These specializations enabled the plants to exploit both the atmospheric and aquatic environments simultaneously.

The dead plant material on the floor of the bog or pond is riddled

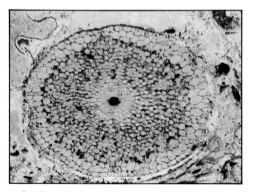

▲ A microscopic cross-section of the fossilized stem of an early vascular plant, *Rhynia*, from Rhynie; this plant was leafless.

through with thread-like fungal hyphae, which is evidence of the decomposers at work. The other inhabitants of this environment were the arthropods. There were tiny water shrimps and primitive wingless insects, the collembolans or springtails. In some of the spore-containing structures, the 'sporangia', minute mites occur which may well have been feeding on the spores. Tiny wounds or lesions can be seen on some of the plants which suggest that some insects were capable of cutting into the plants to reach the sap, rather like greenfly and weevils do today. The Rhynie Chert is another of the fossil record's happy accidents in that an entire environment, albeit a very restricted one, has been preserved in its entirety and can be reconstructed with confidence.

In more extensive freshwater systems, such as rivers and lakes, the commonest remains are of freshwater armoured jawless fishes. These were the first true backboned animals and were detritus-feeders. Arthropods and molluscs (snails and freshwater shellfish) colonized the rivers and lakes. In the past it was firmly believed that the vertebrates first evolved in freshwaters. The main argument for this was based on a study of the vertebrate kidney, which was erroneously thought to have developed in freshwater. A major problem in freshwater conditions is that the concentration of salts is more dilute in freshwater than within living things so that in an attempt to establish an equilibrium, water will enter the animal, which will simply swell up and burst. Water has therefore to be actively pumped out of the animal. One of the chief functions of the vertebrate kidney is to do just this, and it was considered that the development of this facility was directly connected with living in freshwaters. Indeed one zoologist suggested that the development of the fish's bony armour was primarily to serve as a kind of waterproofing. But the armour is spongy and would have been entirely ineffective as a waterproofing agent. It is now realized that certain living vertebrates, such as the jawless scavenging hagfish, which have never had a freshwater stage in their history, do in fact have the type of kidney capable of getting rid of excess water and so the kidney could not have developed as an adaptation to freshwater.

The real evidence that the first vertebrates were sea-living came from Colorado, USA, in the late nineteenth century, when C. D. Walcott described some bony fragments from the armour of jawless fish found in the 500 million-year-old Harding Sandstone. Their broken and worn nature was at first thought to have been a consequence of being washed out to sea from rivers which it was wrongly assumed the fish inhabited. This fossil-bearing deposit is known to extend over many thousands of kilometres, which is consistent with marine conditions. Even earlier fossil vertebrate remains are now known from Upper Cambrian rocks in Wyoming, USA, and these occur in sands deposited in warm shallow seas. The first detritus-feeding vertebrates must have been marine and appear to have evolved from a marine organism like the *Pikaia* from the Cambrian Burgess Shale. The probable existence of a kidney with the potential to get rid of excess water would have made it relatively easy for the first vetebrates to invade freshwater rivers.

Rhyniella

Palaeocharinoides

Protocarus

Rhynia major

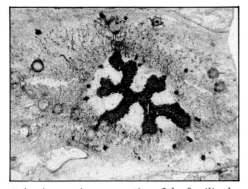

▲ A microscopic cross-section of the fossilized stem of another Rhynie plant, *Asteroxylon*, which bore short leaves.

THE FIRST VERTEBRATES

Jawless fish were the first vertebrates; their fossils are found in freshwater deposits world-wide.

The beginning of the Devonian period, the Age of Fish, started about 400 million years ago, that is, 100 million years after the first record of fossil fish from the Upper Cambrian rocks of Wyoming, USA. The first vertebrates belonged to two contrasting groups represented by *Cephalaspis* and *Pteraspis*. The most obvious difference was apparent when the bony armour was examined in microscopic section. The first such study was by T. H. Huxley (1825–95) in 1858, who recognized that the armour of cephalaspids was constructed of cellular bone, but that of the pteraspids was not. E. R. Lankester (1847–1929) in a monograph on these primitive fishes invented the names 'Osteostraci' (bone shield) for the cephalaspids and 'Heterostraci' (different shield) for the pteraspids. The osteostracans and their associated jawless forms are related to the living lamprey, whereas the heterostracans are related to the stock that gave rise to the jawed vertebrates. Both these groups are found in freshwater deposits in most parts of the world – North America, India, Australia, North Africa and Europe.

The cephalaspid group are characterized by having dorsally placed eyes with a pineal or third eye in between them and a single nasal opening in front of the eyes. One of the most famous studies of any fossils was carried out in 1927, by Prof. E. A. Stensiö, on the cephalaspids of Spitsbergen. It was possible to produce an accurate model of the internal structures of the cephalaspid's head. The connective tissues of the head region were mineralized so that by grinding these

and making accurate drawings of the exposed bone, a magnified scale model in wax could be made. By the time the process was complete the fossil specimen was completely reduced to dust, but it was possible to see in the wax the detailed shape of the brain and the cranial nerves as well as the arteries and veins. This technique was pioneered by Prof. W. J. Sollas of Oxford University, and remains one of the most valuable techniques of the

palaeontologist, although initially it was disparaged. As a consequence of these researches in Spitsbergen, the cephalaspids became the best understood of all fossils. On the basis of the structure of the brain and the nasal apparatus, it was established that there was a close affinity between them and the parasitic lampreys.

The study of the Heterostraci began in the early part of the nineteenth century with two groups

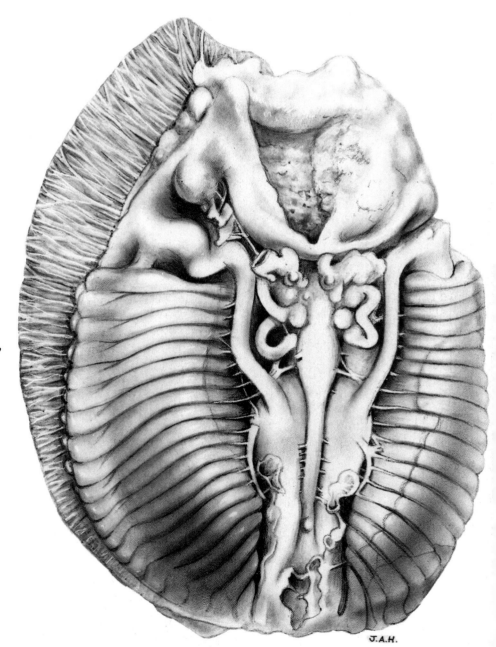

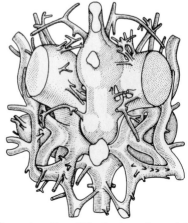

▲ Reconstruction of the brain and cranial nerves of the cephalaspid *Kiaeraspis* from the Lower Devonian of Spitsbergen. The cephalaspids are among the best understood of all fossils.

▲ A drawing of the soft parts including blood vessels and nerves from the head of the galeaspid *Duyunaspis* from the Lower Devonian of south-western China. These were perfectly preserved in haematite 400 million years ago.

being looked at quite independently, the pteraspids in Europe, especially in England, and the psammosteids from Russia. Although in both cases the microscopic structure was studied, it was not until 1899 that it was finally demonstrated that these two groups belonged together. All the heterostracans had a bony armour without bone cells, but they have been divided into different groups according to the pattern of their armour plates. In contrast to the cephalaspids they had a double nasal organ. There were also some forms, the amphiaspids, known from Siberia, where the first gill had been modified to form a spiracle on the dorsal surface which made an inlet duct for the gills, as in skates and rays. This was a specialization for life on a muddy sea-bed, but more significantly represented a major stage in the evolution of jaws, since it is believed that in their formation, the first gill disappears.

There are several groups of jawless vertebrates known and it is clear that they inhabited freshwater rivers and lakes world-wide. In order for such dispersal to have occurred, they must have been able to cross salt water, so it seems likely that there was a marine stage in their life cycle; perhaps like the eel their adult life was spent in freshwater, but their first stages were in the sea.

By the beginning of the twentieth century the main groups of Heterostraci had been described and although many new types have been discovered since then, they all fell into the overall pattern that had already been established. No further surprises were expected, but in 1966 three unique forms, *Galeaspis, Nanpanaspis* and *Polybranchiaspis*, were described from south-west China. The first two were included in the osteostracans and the third in the heterostracans. This view was immediately challenged and a claim made that in fact they represented a completely new group of jawless vertebrates which were named the Galeaspida. The controversy over this continued, until the issue was finally resolved when all the available material was jointly re-examined in 1979. One of the problems was the interpretation of the large dorsal opening which the Chinese scientists claimed was a mouth and the opposition, a nasal sensory organ. The discovery of the collections of oval plates, which fitted into this opening, proved that it could not be a mouth, but some kind of sensory organ.

An important conclusion was drawn from this research: 400 million years ago, south-west China was separated from the rest of the world and the early freshwater-living vertebrates were undergoing their own isolated and independent evolutionary radiation. Among the material collected during the 1970s were a number of specimens, in which replicas of the original soft parts such as venous sinuses, the brain and cranial nerves, were preserved in the iron mineral, haematite. This fossil material provides unequivocal evidence of details of the internal anatomy of these animals. It is evident that the galeaspids are closer to the cephalaspids (osteostracans) than they are to the heterostracans. The new study led to a joint Anglo-Chinese publication in *Nature* in 1979.

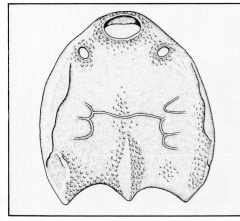

▲ The galeaspid *Polybranchiaspis* showing the controversial central opening near the front of the armour and the orbits.

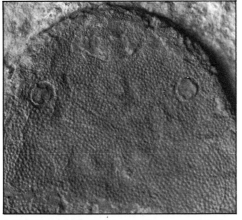

▲ *Polybranchiaspis* showing the cover plate in place on the opening proving that this could not have been a mouth in life.

▲ The cephalaspid *Gylenaspis maceacheni* from the Lower Devonian of Kerrera, Scotland. The sensory regions of the headshield are covered by loose polygonal plates. The eyes can be clearly seen with a single pineal eye between them and a single nasal opening immediately in front.

▲ The heterostracan *Pteraspis rostrata.*

▲ *Cephalaspis lyelli*, the first ostracoderm described in 1835 by Louis Agassiz, from the Lower Devonian of Glamis, Scotland.

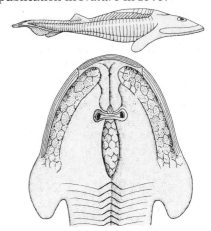

▲ Restoration of the cephalaspid *Hemicyclaspis* in side view and dorsal view of the headshield showing single nasal opening, eyes with pineal eye between, and three sensory fields.

THE ORIGIN OF TOOTHACHE

It has been found that the dentine in modern mammalian teeth is derived from the bony armour of the first vertebrates.

The first vertebrates, the heterostracans, were a major group in the jawless fish, the agnathans; the gill supports had not yet become modified to form jaws, nor had they any teeth. The reason our knowledge of the early vertebrates is so good is that they were covered in a bony armour which was often fossilized. T. H. Huxley noted that the outermost layer of the armour showed fine tubules which had delicate lateral branches with terminal tufts radiating outwards.

The bony armour of the heterostracans had an ornamentation of either ridges or separated tubercles, which are important aids in identifying species. It was the remnants of this bony armour in the mouth region that eventually came to form the teeth of all the jawed vertebrates. Individual tubercles became embedded in the skin rather than fused to the underlying armour. In the sharks and their relatives these same structures remained in the skin as the sharp tooth-like scales. In the higher vertebrates the scales were lost and those associated with the mouth increased in size to survive as teeth. In some instances, such as the tuatara lizard *Sphenodon* of New Zealand, the teeth became fused to the underlying bone; but in crocodiles, dinosaurs and mammals the teeth became embedded in sockets in the bone by means of collagen fibres.

The heterostracans' arrangement of fine tubules in mineralized tissue of calcium phosphate, as noted by Huxley, is identical with that found in the dentine of modern mammalian teeth, except that it is not covered with enamel. So close is it, that competent dental anatomists will confuse a section of fossil heterostracan armour with a preparation of human dentine. In view of the identical structure of the tissue in the fossil armour and in modern mammalian teeth, it is reasonable to infer that the process of formation will have been the same.

Initially the dentine-forming cells line up beneath a layer of the outer skin cells, the epidermis. First they secrete microscopic fibres or 'fibrils' of collagen which are deposited around the fine processes (tiny projections) of each dentine-forming cell. Once this is done the cell produces an enzyme that removes the organic phosphates and allows mineralization to take place (see pp 98–9). The cell with its fine processes then retreats, leaving behind fine tubules and a central thicker tubule with fine lateral branches as evidence of its former position. The process is then repeated and as the cell continues to lay down collagen and mineral, a pattern of converging tubules is formed, marking the line of retreat of the cell. The spaces formerly occupied by the cell processes fill with tissue fluid; in human teeth, they are filled with secondary mineral deposits by the time a person is about 25,

▲ A fragment of heterostracan armour (dentine) showing 'blisters' of new tubercles growing on top of the old surface.

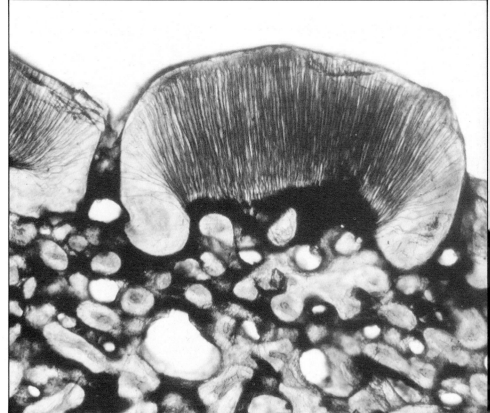

▲ A section of heterostracan armour showing a tubercle with radiating dentine tubules, pulp cavity and underlying bony material.

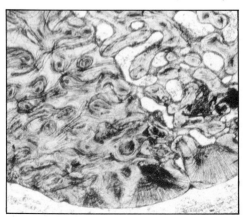

▲ A section of heterostracan armour showing dentine infilling spaces in spongy, bone-like tissue.

which makes them less prone to decay.

The question that arises is what exactly was the original role of dentine? The heterostracans did not have teeth, yet their entire bodies were covered in dentine, which must have had an entirely different function in the jawless vertebrates from that which it has in all the later jawed vertebrates.

Although the dentine initially developed beneath the epidermis, there is evidence from minute scratch marks on the armour made by sand grains that, during life, the dentine formed the outer covering of the animals. As teeth erupt through the gums, so the tubercles erupted through the epidermis, which was sloughed off,

surviving only as a network around the bases of the tubercles.

One of the curious features of human dentine is its sensitivity, yet it does not contain nerves to any appreciable extent. From experiments on volunteers it has been demonstrated that sweet solutions, or hot or cold air, applied to newly exposed dentine will produce pain, whereas solutions known to affect nerves elicit no reaction whatsoever. There is no physiological reason for teeth to have this degree of sensitivity. However, if one considers the possible role of dentine in the first vertebrates, an answer begins to emerge.

Dentine formed the main barrier between the animal and the medium

in which it lived. One of the prerequisites for such a barrier is that it should be sensitive. In modern human teeth anything that affects the narrow tubule of tissue fluid, like the sweetness of sugar, is registered by nerve cells situated in the pulp cavity. Clearly a heterostracan would be unlikely to swim into a sugar solution, but any change in its environment, such as the water becoming more or less salty, would be monitored by this simple mechanism. The structure of dentine reflects its original role as the skin of the first vertebrates.

If this is the fundamental role of dentine then clearly there should be other features of a skin-like nature. By far the most important of these should be a means of regeneration or healing. A study of the fossil armour shows that where pieces have been broken during life, the wound shows the formation of new tubercles sealing the damaged part. This implies that the surviving epidermal tissue around the bases of the tubercles has proliferated, spread across the wound and then induced the formation of further dentine tubercles. If vertebrate teeth are the surviving remnants of the armour of the first vertebrates, then the process of regeneration should be similar. However, in most vertebrates new teeth erupt from beneath. In reality the permanent tooth germs bud off from above as the first teeth erupt; this is similar to the way new tubercles were formed in the first vertebrates' armour. However, the permanent tooth germs move downwards so that prior to eruption they come to be positioned beneath the first teeth.

Many of the processes relating to teeth, which have not previously been fully understood, make sense when seen in the context of the original skin-like tissue. The detailed study of fossils has provided a new perspective to the subject of teeth, and presented for the first time an evolutionary explanation of many phenomena.

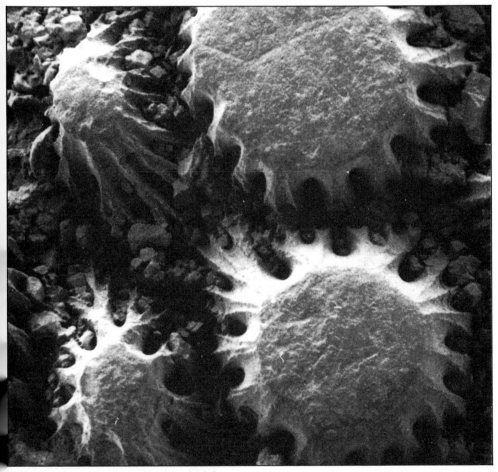

▲ Heterostracan dentine tubercles which have summits showing that wear occurred during life.

▲ The dentine tubercles which ornament heterostracan armour.

▲ Microscopic section of heterostracan dentine, × 500, showing details of tubules with terminal tufts and fine lateral branches.

▲ Microscopic section of human dentine × 500, which is indistinguishable from that of the first vertebrates.

HOW TO RECONSTRUCT AN ANCIENT GEOGRAPHY

The conditions on land and in the sea during the very distant past can be reconstructed by studying the rock sequences and their fossils.

The Old Red Sandstone of the Devonian period was the first sedimentary deposit to contain freshwater and terrestrial fossils. This is the first evidence that life was no longer restricted to the sea, but how these new habitats were taken up is not known. However, these fossils give a fuller picture of past life.

There are several ways of gathering information from the rocks. One of the commonest techniques is to measure the thickness of a rock formation to find out what type of deposit it is. The beds of a marine sediment, for example, will become thinner as the ancient coast is approached and where they thin out and then disappear represents the edge of the sea. A map can be made showing lines joining points of equal bed thickness known as 'isopachytes', just as on an ordinary relief map, contours join up places of equal height. Such maps can be very important

when searching for oil-bearing strata because they provide an accurate record of the thickness and the extent of the accumulated sediments.

Information about ancient environments can also be gained by tracing the geographical extent of rocks of a chosen geological age. For example, if we follow all the Lower Devonian rocks of England and Wales in the Welsh Borderland we find evidence of different ancient environments. There are freshwater deposits of an alluvial flood plain, but in exposures along the coast of the Bristol Channel there are interbedded sands and siltstones characteristic of the influence of tides, and in Devon and Cornwall there are muddy rocks which were deposited under the sea. By crossing this region one is in fact passing from land across an intertidal zone and out to sea.

It is also possible to study the sequence of environments that have

occurred in one place over a period of time. The rock sequence in the Welsh Borderland documents the changes at the end of the Silurian period when the region was covered by the sea, through to the succeeding Lower Devonian when it became dry land. Most of the Welsh Borderland is made up of the Old Red Sandstone. It is an alluvial deposit and compared with a marine sediment fossils are rare; remains of primitive land plants and primitive fishes and arthropods occur sporadically. In the Borderland region, thousands of metres of sediments were deposited, in a sequence of different environments which documents the creation of a new land where formerly the seas had existed. These rocks have been studied in great detail by Professor J. R. L. Allen of Reading University, England. Due to his work we have a great understanding of the geology of the region.

During the latter part of the

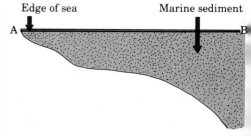

◀ Map of the Upper Old Red Sandstone of South Wales showing lines of equal thicknesses of sediments – 'isopachytes'. The cross-section shows that where the marine rocks end marks the ancient shore.

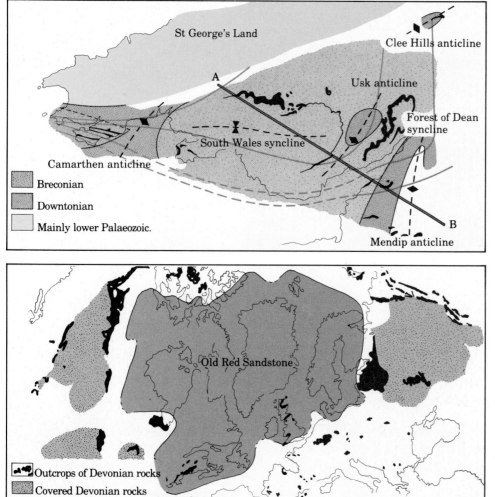

Breconian

Downtonian

Mainly lower Palaeozoic.

▲ Map of the Devonian Old Red Continent. This shows the position of present-day land and sea and where Devonian rocks outcrop at the surface and where they are buried.

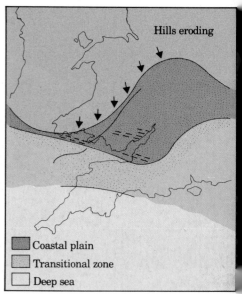

Coastal plain

Transitional zone

Deep sea

▲ Map of Wales and south-west England showing relative positions of land and sea during Devonian times.

Silurian period, the Welsh Borderland was the site of the marine deposition of calcareous silts with limestones containing large brachiopods. Some limestones show cross-bedding and include many shell fragments which means they were subjected to strong currents. These sediments become gradually sandier, suggesting that they are coming closer to the shore. The fossils are of marine invertebrates – more than 60 genera are known and they include brachiopods, corals, bivalves and gastropods. Similar deposits occur today at depths of 15–20m on continental shelves, so it seems evident that the Borderland region was covered by a shallow warm sea into which large rivers emptied, bringing fine sands and muds.

Scattered throughout this succession there are thin bands of bone-beds, comprising worn and broken shells, fish-scales and spines. These are only a few millimetres thick and form thin veneers over scoured and channelled limestones. The bone-beds were formed along a strand line and mark the retreat and subsequent renewed advance of the sea. These were minor fluctuations in the coastline which nevertheless involved the seas retreating to expose many hundreds of square kilometres of sea-bed. Towards the top of this sequence lies the Ludlow Bone-bed, similar to the other bone-beds, but not followed by an advance of the sea. Above this, the sediments are not dissimilar from those beneath the bone-bed, but there are fewer fossils. The mussel *Modiolopsis*, ostracods, small waterflea-like arthropods and the primitive horny brachiopod *Lingula* were relatively common and are indicators of a reduction in salinity, suggesting that the area was covered in brackish waters. These silty beds give way to a thick yellow sandstone with well-sorted grains and ripple marks, cross-bedding, and scour-and-fill channel structures. This sequence of deposition is similar to that found off the seaward margins of major present-day deltas.

The next succession of rocks is dominated by siltstones, at first green but later predominantly red; these siltstones have thin beds of sandstone throughout. Usually the sandstones rest on the scoured and irregularly eroded surfaces of the siltstones, or sometimes on eroded fragments of the siltstones together with drifted plant fragments and pieces of vertebrate and arthropod remains. The siltstones are frequently ripple-marked and show traces of burrowing organisms. The repeated sequences of sandstone and thick siltstones are similar to the succession of sediments preserved in the broad intertidal and subtidal flats along the Dutch and German North Sea coasts and the Gulf of California. The Welsh Borderland seems at that time to have been the site of similar broad intertidal flats in which the soft muds were colonized by burrowing worms and perhaps arthropods. The twice daily sweep of the tides would concentrate the heavier particles, such as pebbles and fish-scales, in the tidal stream channels.

Following the red siltstone succession, coarse sandstones suddenly appear in some places and thereafter a limestone which covers most of the region. This change marks the beginning of the Devonian period and the first fluviatile sediments of the Old Red Sandstone. The limestone is a concretionary rubbly rock with fossil rootlets in it, suggesting it was a soil. These limestones are known as 'calcretes' and form at the present day only on land in hot arid regions. Calcretes are generally found at the

▲ Modern alluvial fans, Sinai, Egypt; and a vertical section of alluvial fan deposits from the Old Red Sandstone.

▲ Quaternary alluvial fan deposits from Norfolk, England.

▲ Section of Old Red Sandstone showing a cross-section of infilled mud cracks from the Welsh Borderland; and a vertical section of overbank deposits and calcrete from the Old Red Sandstone.

▲ Modern mud cracks.

top of a repeated sequence of rock types, beginning with a coarse conglomerate or gravel resting on a scoured and eroded surface. These gravels represent the bottom lag deposits of river channels and as well as coarse sands and pebbles they contain the remains of fishes which inhabited the rivers. Above the conglomerates are thick deposits of sandstones, showing channels, cross-beds and current ripples. These grade up into silts which may show sun-cracks, and this layer is topped by calcretes. In the siltstones there may be traces of crawling and burrowing organisms as well as fragments of land plants. In some fine-grained sandstones complete fishes may be found. All these deposits are typical of alluvial flood plains across which major rivers meander. It was in this environment during the Devonian that land plants and invertebrates briefly flourished when the rivers flooded. Occasionally shoals of fish were swept into small cut-off channels or shallow backwaters where they were trapped and eventually fossilized. Well-preserved fishes are often found in hollows in green siltstones. This colour is due to the presence of land plants which used up oxygen in their decomposition so that the iron present in the soil was not oxidized to form the red ferric oxide which gave the Old Red Sandstone its name. The sequence of gravel, sand and silt represents a single phase of the river breaking its banks.

As the geological sequence is continued, the amount of siltstone becomes less and eventually the rocks comprise simply sandstones and conglomerates. These were laid down by swift-flowing rivers with changing courses closer to the sediment source.

Contained in the riverine rocks are the remains of primitive jawless fishes which are used to identify the geological ages of the rocks. The same species found in the Welsh Borderland are also found in Nova Scotia and even in Spitsbergen. The identity of many of the species suggests that they lived in the same region, and is strong evidence that they were not separated by great distances as their fossils are today.

The fishes that inhabited the rivers and lakes seem to be confined to rocks laid down in these freshwater environments and not in marine sediments. However, the overall distribution of these fishes can only be explained if they were able at some time in their life-history to live in the sea. As they seem to have lived in freshwaters for most of their lives, they may have spent the early stages in the sea, the environment in which they

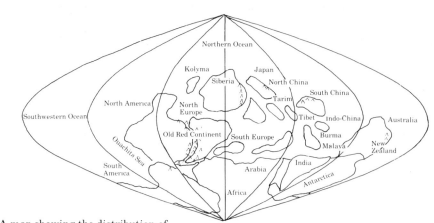

▲ A map showing the distribution of continental masses during Devonian times.

▲ Lower Old Red Sandstone intertidal and alluvial sediments at Lydney, Gloucestershire, England; and a vertical cross-section of the repeated sequence of rocks at Lydney.

▲ Small-scale ripple-bedding preserved in Quaternary intertidal marine sands at Lowestoft, Suffolk, England.

116

▲ Modern intertidal sand showing perfectly symmetrical ripples.

▲ Modern intertidal rippled sands.

▲ Offshore sand shoal, North Sea coast; and a vertical section of a Lower Devonian offshore marine shoal from Downton Castle, Shropshire, England.

▲ Lower Devonian calcretes, Pembrokeshire, South Wales.

first evolved, and in this regard must be thought of as more akin to the eel than, for example, the salmon.

From the direction of the cross-bedding in the Old Red Sandstone of the Welsh Borderland it was evident that there must have been a source to the north and west from which many thousands of metres of sediment were derived. The existence of a giant mountain chain was postulated on an unknown continent where the Atlantic Ocean lies today. It is now known that there was indeed a mountain chain which was produced when what are now eastern North America and western Europe collided to form the Old Red Continent, eliminating the ancient Iapetus Ocean (see pp 190–1).

Information about the climate in the Devonian period has come from studies of calcrete limestone. The presence of Devonian calcrete suggests that the climatic regime of the Old Red Continent was similar to that of the regions where calcrete forms today. These regions are situated between latitudes 20° and 30° and have a seasonal rainfall of which the monthly maximum is less than a quarter of the annual rainfall. It can be deduced from these observations that the climate in the Devonian was arid and subject to seasonal flooding with a mean annual rainfall of 100–500 mm and a mean annual temperature of 16°–20°C.

We can see evidence in the sediments of different ages that the climate changed during the passage of time. The Devonian was succeeded by the Carboniferous period, which was dominated by humid tropical swamps and forests; and the later Permian and Triassic periods were notable for the return of the hot desert conditions of the New Red Sandstone. This could have been achieved in two ways: there could have been a shift in the positions of the climatic zones world-wide so that with overall heating or cooling the zones migrated. This certainly happened during the recent Pleistocene ice ages of the past two million years as the ice-caps advanced and retreated. The other possibility is that the continents themselves shifted their positions relative to the climatic zones. The Old Red Continent may have drifted from the savannah and hot deserts of the southern hemisphere across into the humid tropics during the Carboniferous and on to the hot desert zone of the northern hemisphere during the Permian and Triassic. Whether it was the climatic zones or continents which moved, the results would have been the same. It was not possible to decide either way until the discoveries of the 1960s (see pp 190–1).

LAND PLANTS

**The land-living species
of plants evolved
from the aquatic plants
as they reduced their
dependence on
water for reproduction.**

By the end of the Devonian period,
that is within the space of 100 million
years, plant life had spread from the
waterlogged edges of freshwater rivers
and lakes to cover vast expanses of the
surface of the land. This happened
because plants gradually developed
more efficient reproductive processes
which became less and less dependent
on water.

The surfaces of the continents were
transformed – instead of being made
up of bare rocks they were now clothed
by plant life. The entire cycle of erosion
was changed by land plants. They
introduced a new process of chemical
breakdown of rocks by decomposing as
well as physically breaking up rock
with their roots, and for the first time
soil was formed (see pp 14–15).

The very first sign of the existence
of land plants comes from Silurian
rocks some 430 million years old.
These comprise microscopic spores
which must have been windborne and
are therefore assumed to come from
land plants. The first direct evidence of
land plants comes in the form of
fossilized stems and branches from
later Silurian rocks in Wales. These
are from the plant *Cooksonia* which
had no leaves, simply a stem which at
the tip carried capsules containing
spores; it was one of the first plants
with roots for drawing water, not
merely for clinging. The importance of
Cooksonia is that it had a system of
water-conducting cells of xylem or
wood in which the original cells had
died and their walls strengthened to
act as vessels for conducting water. For
this reason, these and all such plants
are known as vascular plants.

We can observe primitive vascular
plants such as ferns, club mosses and
horsetails on land and they all produce
spores in order to reproduce. The
spores, however, can only develop in
humid conditions where they grow into
a thin sheet of plant tissue, the
'prothallus', which produces male and
female reproductive cells. The male
'sperm' are released into water in the
soil and they swim until they meet a
female egg from another prothallus.
When this is fertilized, it grows into a
new fern or horsetail. Thus there are
two separate stages in the life cycle.

This method of reproduction which
must take place in water, even if only
within damp soil, restricted the
environment that was capable of being
colonized by such plants. The major
change that took place 350 million
years ago was when sexual generation
took place on the parent plant. The
first indication that this stage had been
reached comes from fossil spores.
Instead of the spores being released
and then developing into prothalli
which produce sperm and eggs, two
different types of spore were produced:
large spores or megaspores, which
remained attached to the parent plant,
and microspores or pollen grains that

were shed into the air. The megaspore
consisted of female spores in a spore-
case partially enclosed in a cone-like
protective covering, which together
made up the ovule. The microspores
were blown onto the opening of the
megaspore and the male sporangium,
or spore-case, is released in a droplet of
water. The drop of water with the
sporangium makes its way down inside
the megaspore and then forms a sperm
cell; the female sporangium breaks
open to expose the female egg cells.
Fertilization takes place and the ovule
becomes a seed. The 'integument', or
covering, dries and hardens and the
protected, developing, plant embryos,

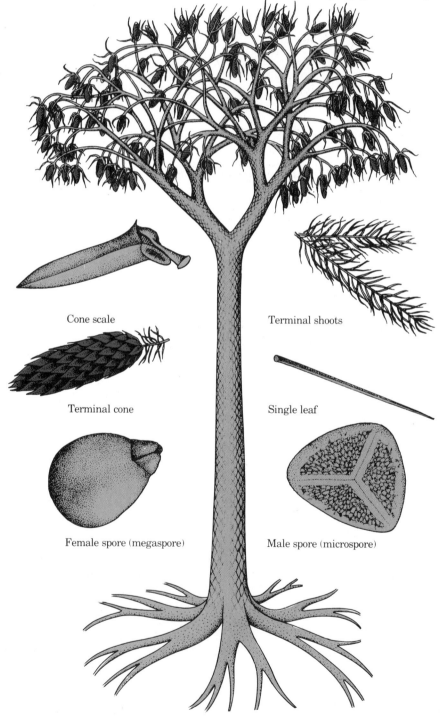

Cone scale

Terminal shoots

Terminal cone

Single leaf

Female spore (megaspore)

Male spore (microspore)

▲ A reconstruction of the 300 million-year-old
giant club-moss *Lepidodendron* showing the
different parts of the plant which have been
found as fossils, enabling a complete picture
to be built up.

the seeds, are released from the parent plant; if they fall onto a suitable environment they will rapidly germinate and produce a new plant.

Once this method of reproduction had become established in the Devonian period, the way was open for plants fully to colonize the land. This stage of plant evolution survives today in the 'gymnosperms' or 'naked seed' plants, such as the maidenhair tree (ginkgo) and the cycads. Unlike the cycads, the pines and fir trees produce pollen and egg-bearing cones on the same tree. What is more, the male cell does not need a drop of water to help fertilize the egg. The very beginning of these gymnosperms is represented by the Upper Devonian fossil *Archaeosperma* or 'ancient seed'.

There was, however, one mechanical problem to be overcome. A system of woody vascular tubes with growth taking place at the tip of the plant puts a restriction on the height to which a plant can grow and still support itself. There needs to be some means of increasing the girth of the plant. This was accomplished by the development of secondary thickening by a layer of cells beneath the bark called the vascular cambium which divides to give new wood. In the growing season in the spring when large amounts of water are required, the newly formed wood has large cell spaces; when growth is reduced in winter, the cells are smaller and thicker walled. In areas where there are differing seasons, say a wet and dry, this difference in growth pattern will show up as rings. In fact the Devonian *Archaeopteris* shows just such rings, which is evidence of there having been distinct seasons at that time. This particular plant had a trunk with a diameter of a metre and its height has been calculated as 30m. Indeed, by the end of the Devonian period there was a proliferation of large forest trees.

Plant life had become firmly established on land during the Devonian period, as most of the major stages in plant evolution had occurred. The next major evolutionary step took place almost 200 million years later during the Cretaceous period (about 100 million years ago) towards the latter part of the Age of Dinosaurs. This was the beginning of the flowering plants or 'angiosperms', which developed a more sophisticated method of reproduction. The flower is the reproductive organ of the plant. To prevent self-pollination, both male and female parts of the flower develop at different times. The flowers are built up of coloured modified leaves, the petals. These are to attract insects which pollinate the flowers. The reproductive organs are close together for easier pollination and comprise stamens which shed pollen grains and a female stigma, or receptive spike, on top of the ovaries. Pollen grains are carried by wind or other agents, but in particular by insects which give wide distribution. Once the pollen reaches the stigma, a tube grows down from the pollen until the egg is fertilized. Inside the egg is a seven-celled prothallus; two male nuclei carried into the pollen tube fuse with nuclei of the prothallus – one fused pair grows into the embryo and the other forms food reserves. The seed matures and is dispersed when the ovary wall splits.

The method of reproduction and the variety of mechanisms for seed dispersal have made the flowering plants astonishingly successful. There is little doubt that angiosperms originated from gymnosperms, but to date there is no clear evidence as to exactly which gymnosperm group is the ancestor. One of the more recent discoveries made in 1981 in Scania, Sweden, was the preservation of complete fossil flowers which confirmed the existence, during the Cretaceous period, of all the parts of the flower that had previously only been inferred.

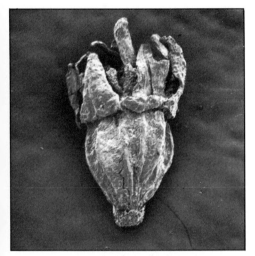

▲ The first evidence of flowering plants – a fossil flower preserved in Cretaceous mud in southern Sweden; it measures about 2 × 1 mm.

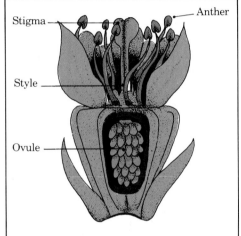

▲ Diagram of the fossil flower which has been preserved in a three-dimensional state, showing part of the ovary wall removed.

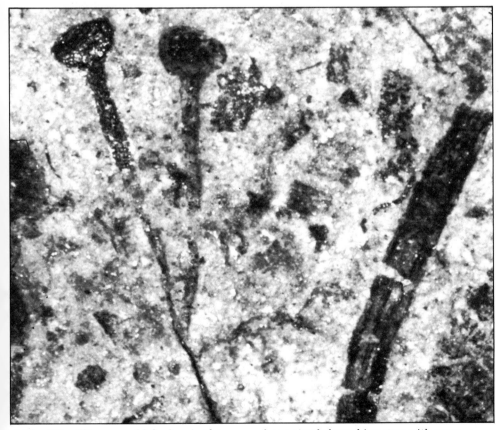

▲ The 410 million-year-old *Cooksonia*, the very first vascular plant in the fossil record, from Upper Silurian rocks, South Wales. It has a simple branching stem with spore capsules at the tips, but no leaves.

LAND ANIMALS

The development of the closed egg freed the vertebrates from living near water, and animals, like plants, began to colonize the land.

The conquest of the land by animals parallels that of the plants that preceded them. The similarities are particularly striking and, as with the plants, many of the major stages in this evolution have survived to the present day and are able to provide a valuable insight when viewed in conjunction with the fossils.

During the Devonian period, the Age of Fish, there developed a wide range of different groups: the heavily armoured placoderms gave rise to the modern cartilaginous fishes such as the chimaeras or ratfish, and the sharks and rays; the acanthodians, the so-called 'spiny sharks', represented the basic stock of all the later bony fishes. In both lineages their paired fins served primarily as stabilizing organs preventing undue roll. As they evolved, the base of the fins became narrower so that a greater range of movement became possible and this served as an important aid for increasing manoeuvrability. The internal skeletal support of the fins developed in two contrasting directions. In some, the

skeletal support was concentrated in the main axis of the fin and was known as the lobe fin; in others, the skeletal elements radiated out from a narrow base near the body wall, and were known as ray fins. Both basic types of fin were found in the cartilaginous and the bony fishes.

From geological evidence, it is apparent that the land through which the rivers flowed and in which the fishes lived was subjected to seasonal droughts and floods. The fishes, in many ways, found themselves in the same situation as the early freshwater plants. Outside of the medium which they both normally inhabited were substances which they required: carbon dioxide for the plants and oxygen for the fishes. This was especially important for fishes during the seasonal drought, when rivers would dry up or be reduced to isolated pools in which all the oxygen would be used up and the ponds become fouled. Clearly any fish capable of exploiting the oxygen in the atmosphere would

stand a far greater chance of surviving; most later Devonian fishes developed some kind of accessory breathing organ, a kind of pouch or lung from the gill region. The living lungfish are survivors of the fish of that time and can supplement their oxygen requirements by breathing air. In all modern bony fish, for instance the herring family, this primitive lung has taken on a new role, that of a hydrostatic or buoyancy organ, the so-called air-bladder.

The second feature that developed, once again to ensure survival during crisis periods in the year, were strong muscular fins. These were present in both the primitive ray-finned and the lobe-finned fishes and enabled them to crawl through mud in search of prey and a better environment, if their own pool became fouled or dried out.

Both air-breathing and movement on land are still characteristic of living fish and in neither the Devonian nor present-day fish are these characteristics related directly to

▼ Some of the first egg-laying land animals which inhabited the Carboniferous swamps. In the centre foreground is a small aquatic or semi-aquatic amphibian *Microbrachis*, which has greatly reduced limbs. Behind is *Urocordylus*, another amphibian. On the far right is *Petrolacosaurus*, which is probably an ancestor of many later reptiles. The squat *Ophiacodont* is behind, a member of the line which led to the paramammals and eventually the mammals. The plants are club-mosses, tree and seed ferns and horsetails.

colonizing the land, but merely to ensure success as fish. The first animals that were capable of breathing air and walking on land are known as amphibians, that is, dual animals. Although they occurred in the Devonian, it was not until the succeeding Carboniferous period with the spread of shallow swamps, lush vegetation and a large array of arthropod life that dual animals, such as salamanders and frogs, came into their own. Exactly like the first land plants, the larva stages of development were spent exclusively in water. The tadpoles of these amphibians were fish-like, being gill-breathers. Like the ferns, the amphibians were of necessity restricted to living close to water for reproduction.

The real advance in freeing the vertebrates from complete dependence on water came with the origin of the 'amniote' or 'cleidoic' (closed) egg in the Carboniferous era. This egg instead of being deposited in water, as with most modern amphibians, was retained within the mother, the sperm being emplaced by the male. After fertilization the embryo was provided with a food supply, the yolk, and was enclosed in a fluid-filled bag, the 'amnion' (thus all higher vertebrates are called amniotes). A membrane, the 'allantois', was concerned with respiration and the excretion of waste products, and there was a protective covering around which a shell of calcium carbonate could be deposited.

The development of this type of egg was one of the major evolutionary events in the history of the vertebrates, enabling them to colonize dry land, even though the egg was clearly not invented for such a purpose. If we examine it in the context of the period in which it occurred and the conditions under which it developed, we find another answer.

It is evident that this method of reproduction occurs in both living birds and mammals, as well as reptiles, which suggests that the major division between the ancestors of the birds and the mammals must have happened after the amniote egg had originated. It is now firmly established from details of the structure of the skulls that these two lineages were already separate within the Carboniferous period, and it was not until the very end of the Carboniferous that the vertebrates ventured onto really dry land. It can only be concluded that the amniote egg was invented many millions of years earlier in the Carboniferous, before it was really needed for the conquest of dry land. This indicates that it fulfilled some quite different function that was advantageous within the swamps.

Laying large numbers of eggs in open water, as frogs do, results in the provision of food for numerous other creatures. Even when the tadpoles emerge they are similarly vulnerable to any predatory insect larva. There is, therefore, an advantage to be gained by not laying eggs in a densely populated environment. Indeed, modern amphibians go to inordinate lengths to avoid this with a whole series of bizarre specializations. Hiding on land a few eggs which have a sufficient food supply and a means of breathing and avoiding dessication would seem to be an extremely economic, efficient reproductive strategy for a swamp-dwelling animal, like the tropical frog. Although this reason for egg development is speculative, it makes sense in terms of the period and conditions in which it took place.

▲ The amniote egg which was developed in the Carboniferous period.

COAL SWAMPS AND DELTAS

During the humid tropical climate of the Carboniferous period, the forests decayed and formed peat which eventually became today's coal deposits.

The Carboniferous rocks show that 340 million years ago the northern hemisphere was covered by vast swamps, where dense, humid forests grew. (There was a large ice-cap centred on the southern continents.) It was the northern forests which were responsible for much of today's world coal. As the plants and trees decayed under varying compression, different types of coal were formed.

Coal formation begins with partially decayed plant matter accumulating in tropical swamps. As it is buried by more and more plant debris, so fungi and bacteria are prevented from completely breaking it down. Through a process of compaction and slight heating, the accumulated plant debris converts to coal in a series of stages. First lignite or brown coal is formed, then bituminous coal, and if this is subjected to the enormous pressures associated with mountain building, anthracite (a hardened form of almost pure carbon) is produced. These stages mark a progression of increasing carbon content with decreasing gas and water content.

Coal is generally shiny when freshly broken, but there are often dull-looking layers known as 'fusain'. This is dusty, exceedingly brittle and fragile. When this part of coal is examined under the scanning electron microscope, it can be seen that the cells of woody tissue are perfectly preserved. This material is fossil charcoal and was formed in exactly the same way as newly burnt charcoal. This occurs when a piece of wood burns in insufficient oxygen so that combustion is slowed down and prevented from completion, resulting in charcoal which is virtually pure carbon. This product, halfway to becoming coal, can serve as an efficient fuel because, although much of the water has already been driven out, the carbon has not combined with oxygen to form either carbon monoxide or carbon dioxide. Charcoal forms naturally during forest fires and the presence of fusain in coal provides proof of forest fires occurring over 300 million years ago.

The presence of insects and non-marine bivalves in the coal seams confirms that freshwater was associated with the land where coal was being formed. We know that coal was formed in a terrestrial environment because easily recognizable plant fossils are commonly found when coal is split. These can be pieces of bark, or branches and leaves set in a fine-grained matrix made up of broken-down plant matter like seed and spore coverings, leaf cuticles, tannins and resins. The lack of oxygen in the peat from which coal is formed allows this plant matter to be preserved.

Coal occurs in seams of variable thickness and is always associated with a characteristic sequence of rocks; in fact the coal stratum marks the termination of a regularly repeated cycle which is not always complete – it may have parts missing or abbreviated. However it is only when the repeated

▲ An *Alethopteris* plant frond fossilized in coal.

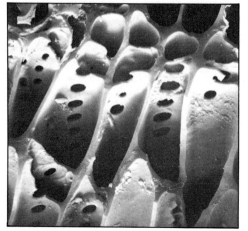

▲ Jurassic fusain from gymosperm wood.

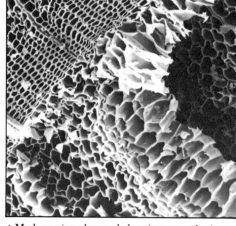

▲ Modern pine charcoal showing growth ring.

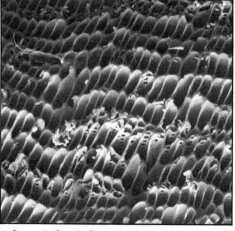

▲ Detail of the Jurassic fusain.

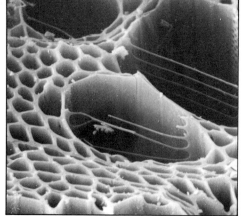

▲ Modern beech charcoal.

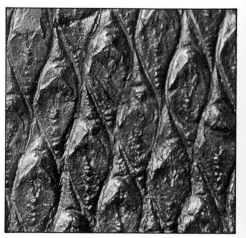

▲ A coal fossil of the bark of *Lepidodendron* showing leaf scars.

sequence of rock types is studied that the wider environmental context of the coal swamps can be understood. The presence of marine and freshwater rock sequences is evidence that this environment was close to sea-level in coastal swamps like those that occur in the humid tropics today.

The cycle is a record of a deltaic environment and begins with a limestone or shale containing marine fossils, which represents offshore shallow marine conditions where very little sediment from the land was being deposited. The first signs of sediments derived from the land are shales containing freshwater mussels and clays and muds. This sequence of rock types is simply a record of where the rivers dumped their sediment in the sea. The coarser material was deposited first and the finer silts and muds were carried further out. The

▼ A vertical section of rock strata showing the repeated sequence of rock types found in the coal measures.

Shale
Limestone

Coal
Seat earth
Sandstone with large-scale current bedding

Sandstone with small-scale cross-lamination
Silts and shales

Marine shale

Marine limestone

successive rocks in the sequence become coarser and more sandy with small-scale cross-bedding followed by large-scale cross-bedding. These rocks show that as more and more sediment was deposited the margin of the sediments was pushed further and further out to sea to form a delta.

The fine sands and muds were spread over the sea-floor at the foot of the delta and because of the fresh water associated with the influx of sediment, the delta's seaward margin was able to support a fauna of freshwater or brackish animals. But in front of this area, the heaps of sand tipped over and the delta slope moved forward, leaving little sign of life in the accumulated sediment.

The next layer in the sequence is a fine-grained pure sandstone or 'gannister' with rootlets preserved in it. (Gannister is used in foundries for its refractory properties.) This rock represents the rootlet beds of the Carboniferous forests which removed all the nutrient to leave a pure sandstone. As the large sluggish rivers which built up the delta approached the sea, they divided into separate channels and a network of creeks. Between these lay the marshes and swamps and it was here that the Carboniferous forests became

established. The coal seam which ends the rock cycle shows that in these waterlogged conditions the trees became peat when they died and eventually formed the coal. Where channels in the delta became cut off they formed lagoons which in time were filled by sediment and recolonized by plants. These conditions tend to be characterized by finer sediments which are usually flat-bedded.

A single rock cycle can be explained as the result of a delta pushing out into the sea. The rocks also record the existence of repeated cycles and deltaic deposits thousands of metres in thickness. In modern deltas like the Mississippi, the river breaks its banks about every 500 years and begins building up the delta in a new direction. Along the deserted course the sediments are then inundated by the sea and a marine-dwelling fauna becomes established. The whole cycle gradually builds up again. This process, however, cannot explain the enormous depth of such repeated cycles, it merely explains the mechanism of the transition from one cycle to another. From examining both the coal of the swamps and the sands of the delta, it is clear that the entire pile of sediments has subsided, presumably under its own weight.

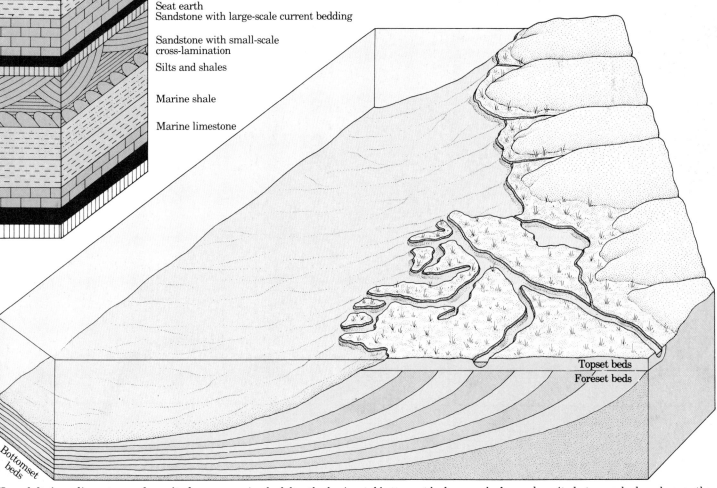

Topset beds
Foreset beds

Bottomset beds

How deltaic sediments are deposited
▲ Three main types of deposit can be found in a simple delta: the horizontal bottomset beds are deposited on the sea-bottom; the foreset beds are deposited at a marked angle over the bottomset beds; topset beds lie over the foreset.

THE EVOLUTION OF FOOD WEBS ON LAND

Food webs demonstrate the relationship between animals and plants in a given environment. As environments and species evolve, these interrelationships alter.

The interdependence of all living things is reflected in the food web. Over time, as climates have varied and environments changed, new plants and animals have developed and the relationships between them have evolved and altered.

The observation of modern food webs can act as a guide to understanding the ecosystems of the past. The evolution of species within ecosystems causes patterns to change in the number of species within a given area, the ratio of these species to one another, and the structure of the food web. Modern food webs demonstrate how these relationships vary from one environment to another. Which animal fills what ecological niche is determined by the size of a land area and its climate. The continent of Africa, for example, would obviously have a greater variety of niches and animals than a small island.

What is a terrestrial food web? The consequence of plants utilizing water and minerals in the soil and, with the energy of sunlight, carbon dioxide in the atmosphere, was that the land became green. Plants produce complex organic molecules such as sugars and they are therefore known as primary producers. Plants form the basis of all life on Earth and are consumed by herbivorous animals, the primary consumers, which in their turn are devoured by flesh-eaters, the secondary consumers. The primary consumers utilize only some 20% of plant material and of that only about 10% is stored to become available for the secondary consumers, the carnivores. The latter may themselves be preyed upon by tertiary consumers, the 'top' carnivores. At each stage in this chain the amount of energy lost or wasted is enormous, so there is a relatively small number of top carnivores, compared with herbivores. At the present day, when 100m² supports a herbivore, a primary carnivore would require 1,000m² whilst a top carnivore would require 10,000m². This is the basic ratio governing the relationships of living creatures to one another.

It is possible to work out the most

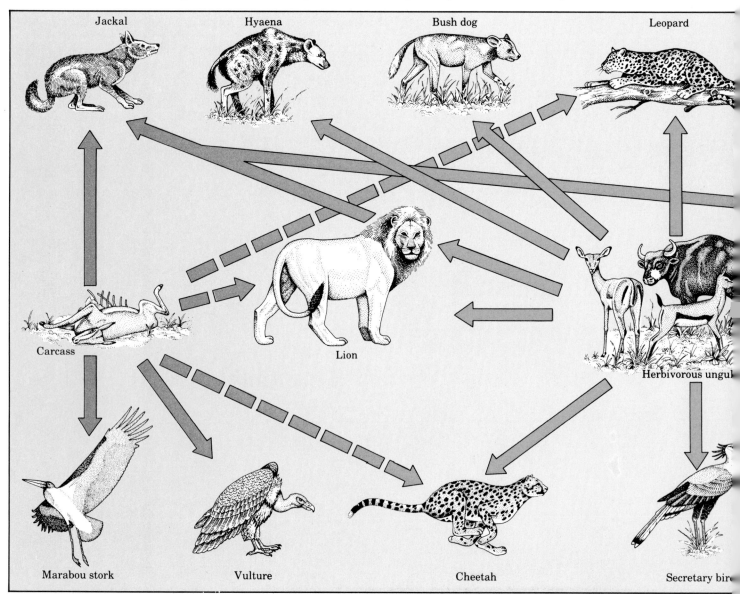

A modern food web
In a modern environment the majority of animals (95%) feed directly on plant material. They are preyed upon by a small number of carnivores (about 3%). Within this food web there exists a more primitive pattern which is

likely relationships of animals in a simplified type of food web. If any patch of ground on which plants are growing is investigated, there will be evidence of parts of the plant being consumed by slugs and snails, or insects such as aphids (greenfly), or insect larvae such as caterpillars. Leaves and other parts of the plant, such as petals and fruits, fall to the ground and are broken down by the action of fungi and bacteria, the decomposers. Partially decomposed plant material, especially leaf litter, is consumed by a variety of animals, the most familiar being earthworms, the detritivores. It is worth noting that in 1881, a year before his death, Charles Darwin published his book *Formation of Vegetable Mould through the Action of Worms* which was a major step in the recognition of the importance of earthworms in maintaining the fertility of the soil. At the time, the attitude towards earthworms was that they were noxious animals which should be destroyed. Indeed, the soil

scientists of Darwin's day were highly critical of the scientific value of Darwin's contribution. However, subsequent research has vindicated Darwin.

It is possible to extract a good variety of minute organisms from the soil: an enormous number of worms and minute arthropods, woodlice, millipedes and centipedes, which can be found by lifting up stones and examining the life around plant roots and digging in the soil. All these animals and the decomposers, fungi and bacteria, are the primary consumers. Hunting the slugs and snails, worms and insects are small mammals such as shrews and hedgehogs. These furry or spiky mammals are secondary consumers and they in turn are preyed upon by owls and bigger carnivorous mammals such as weasels, stoats, foxes and cats, which are the tertiary consumers.

This seems to be the fundamental pattern of terrestrial food webs which was first established during the Carboniferous. In fact even this simple

case is complicated at the present day by a further strand: there are mammals, such as mice and squirrels, that feed directly on plants and are then preyed upon by the top carnivores, the owls, stoats, cats, etc. The energy route to the top carnivores from plants can be via invertebrates, which are preyed upon by insectivores, or via larger animals which can feed directly on plant matter. This particular type of food web comes to its logical conclusion on the plains of East Africa. There large numbers of mammals feeding on plants are being preyed upon by a relatively small number of carnivores. The primitive plant-litter based food web continues to exist but is overshadowed by the more dramatic game animals and their means of survival.

The modern food webs or food chains are fairly well understood, but in 1980, research was done into the first type of food web that developed in the Carboniferous period with the first land animals, the amphibians. This

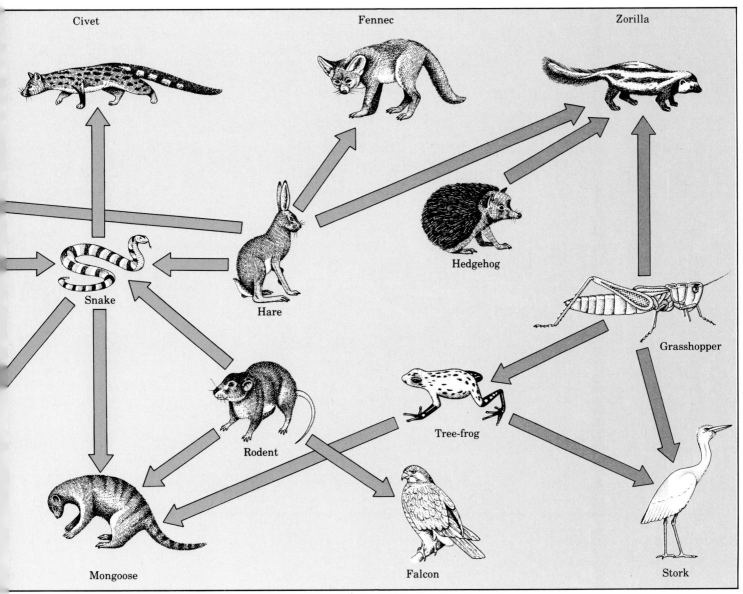

based on insects feeding on plant material. Other components of the food web are the scavengers feeding on animal remains and general detritivores which feed on the decaying remains of other organisms.

kind of work involves studying all the different kinds of fossil remains in the context of the environmental conditions which can be ascertained from a detailed examination of the rocks. Three major types of environment in the conditions of the coal swamps can be recognized, which were characterized by different kinds of amphibian. In the open water of extensive lakes or blocked river channels were many different kinds of fish – the acanthodians, primitive ray-finned fishes and shellfish-eating lungfish; preying on these were large crocodile-like amphibians over 3m long with reduced limbs. Within the swamps there were two types of environment. The shallow waters contained many fully aquatic creatures with reduced limbs and elongated swimming tails. Some resembled sea-snakes, others were newt-like, many of the smaller forms having external gills. Where the vegetation was prolific, there were more terrestrial animals which spent much of their adult lives on the drier land, although the tadpoles still lived in water. Some of these aquatic amphibians had lost their limbs, others looked like salamanders. Perhaps the most important animals were the first reptiles, the romeriids, and the primitive paramammals (the branch of the reptile family which led to

mammals), the ophiacodonts, which lived in the tropical swamps but laid their eggs on dry land.

One of the features of the coal forests was their very tall trees with leaves forming a high forest canopy. Insects and other invertebrates were capable of exploiting this source of food. By this period there were many kinds of flying insects, such as dragonflies and stone-flies, and preying on them were web-spinning spiders. Both spiders and insects became the food of the smaller amphibians.

The leaf litter on the forest floor and organic debris in the shallow waters provided the source of food for most of the animals. In the waters were various arthropods, gastropods and aquatic worms and insect larvae. In the drier parts, arthropods, such as cockroaches and millipedes, as well as snails fed on the litter. The aquatic invertebrates provided food for the tadpoles of most of the amphibians, as well as the fully aquatic creatures. The water-living amphibians and tadpoles were preyed upon by the larger amphibians, just as in the drier parts of the swamps the smaller insect-eating amphibians were preyed upon by the larger.

The wide range of different amphibians and the new reptiles was part of a food web founded on leaf litter. There is no evidence that any

vertebrate was capable of feeding on plant material directly; all the vertebrates seem to have been meat-eaters, even though the smaller ones only ate insects and worms. In this respect the Carboniferous food web represented a type that was totally different from those of the present day.

Towards the end of the Carboniferous period many of the coal swamps dried up; the lush tropical forests gave way to much hardier forms of plant life, such as conifers. Similarly the majority of the amphibians died out with the demise of their environment. It was at this stage that the animals capable of laying shelled eggs on dry land not only found themselves with a means of survival but were capable of fending for themselves in more arid regions so long as there was some supporting plant life for a food web to become established. It was from this point on that the real conquest of the land by animal life was first established.

It was not until the early 1960s that scientists began to be aware that the evolution of communities was outlined in the fossil record. This stemmed from the research in Texas and Russia by Everett Olsen on early paramammals. The reptiles were specialized for eating meat; there was no terrestrial vertebrate that could be unequivocally designated a herbivore.

▼ The first terrestrial food web, which developed during the Carboniferous period, was very different from the present day because there were no plant-eating vertebrates. At the base of the food web were invertebrates – insects, arthropods, gastropods and worms feeding on plants and organic debris. The diagram shows how the primitive vertebrates – ranging from aquatic animals to paramammals – preyed upon each other.

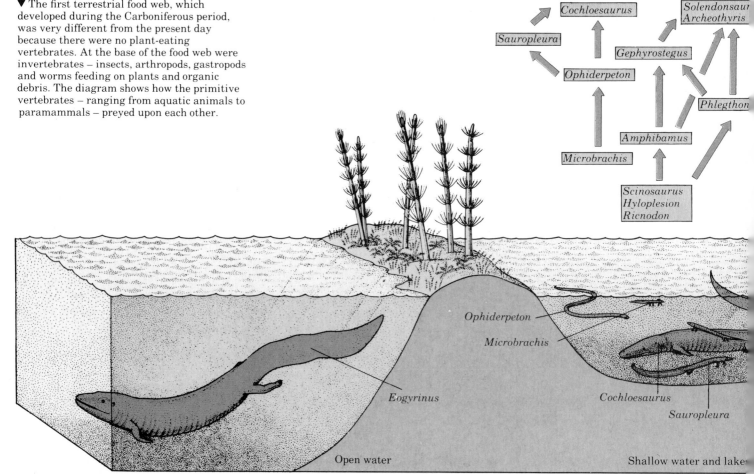

As it was assumed that the food web of land animals would have revealed a preponderance of herbivores upon which a relatively few carnivores would have preyed, as in the modern food web, the fact that all the faunas seemed to be exclusively carnivorous seemed most unusual. The initial reaction was to believe that this was due to the vagaries of the fossil record and that the herbivores were not preserved in that particular region. However, work on the Permian period began to demonstrate that at different geological ages there seemed to be changes in the proportions of the different types of paramammals in existence. Fossils from rocks of the same age in Russia showed the same pattern. Exclusively carnivorous faunas seemed to be part of a system.

The Permian period is the most important of all in one respect. At the beginning of this period, the type of food web was established where none of the vertebrates was capable of feeding directly on plants. Yet at the close of the period and at the start of the succeeding Triassic period, a completely modern type of food web had become established. In both North America and Russia it is possible to work out a variety of food webs which evolved; it appears as though different experiments of community evolution had taken place until the optimum arrangement emerged.

At the beginning of the Permian, in spite of the developing aridity, there was still a food web dominated by an aquatic phase. Plants either directly or via organic plant debris and leaf litter were consumed by invertebrates which were eaten by smaller reptiles, which in turn were preyed upon by larger reptiles; in fact the largest of the carnivorous reptiles seemed to have preyed on one another. However, most of the invertebrates were fully aquatic and these were eaten by fishes or amphibians which were finally devoured by semi-aquatic reptiles. Later in the Permian of North America the food web was entirely terrestrial. There seemed no longer to be any substantial aquatic contribution. Moreover, a proportion of the vertebrates were herbivores and were becoming equal in importance to the invertebrate feeders. In fact both the invertebrates and the plant-eating vertebrates contributed equally to the sustenance of the large flesh-eaters.

In the Upper Permian of Russia a small localized type of food web developed that was based on invertebrate feeders, as there were no herbivorous vertebrates, although they were known in abundance in rocks of the same age elsewhere. Just as in a modern woodland a food web can be constructed which is a surviving example of the Carboniferous plant-litter-based web, so this later Russian fauna represented not a reversion to a primitive type of community, but simply its survival.

At the very end of the Permian in both Russia and North America there had evolved a terrestrial food web in which the majority of vertebrates were herbivores, and the invertebrate feeders had become a minor part of the fauna. The big carnivores were supported by a large population of herbivores, although they were sufficiently numerous to have preyed on one another as well.

The beginning of the succeeding Triassic period showed that in all essentials the same type of food web had continued but that the proportion of invertebrate feeders had increased. At this stage, there was a significant change in the ratio of carnivores to herbivores: the carnivores made up a very small proportion of all the fauna – about 3%, which is the same percentage as found today. The Permian period witnessed the gradual evolution of the modern food web. This was in part due to the reptiles – paramammals which were evolving gradually into mammals. As the Triassic period unfolded, this modern ecosystem collapsed and the Earth was dominated by a different group of animals: the dinosaurs.

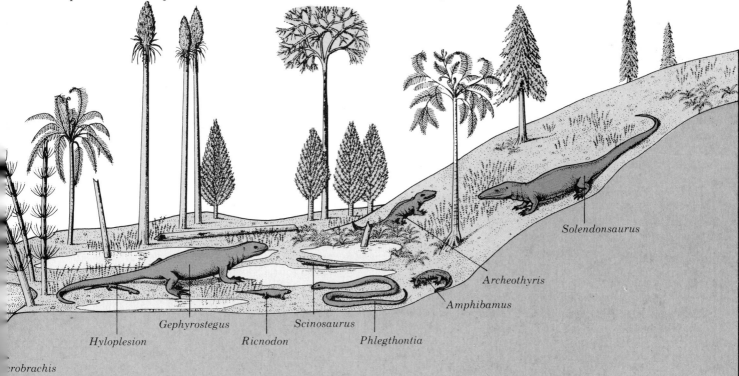

Hyloplesion   Gephyrostegus   Ricnodon   Scinosaurus   Phlegthontia   Amphibamus   Archeothyris   Solendonsaurus

robrachis

Swamp   Upland

THE EMERGENCE OF THE DINOSAUR

Dinosaurs were reptiles that emerged from water onto land. Their success as land predators originated ironically in their ancestors' adaptations to living in water.

Animal life on land was dominated during the Permian, for some 70 million years, by the paramammals or mammal-like reptiles. They established the modern type of food web in which a small proportion of carnivores is supported by a large herbivorous population. Then towards the end of the Triassic period, about 200 million years ago, the dinosaurs took over and thereafter, for 140 million years, dominated the world. This much is evident from the fossil record, but we need to ask how this new group of animals entered into a firmly established ecosystem and became dominant.

When the fossil record is examined in more detail, it becomes possible to work out how this change must have come about. The dinosaurs' ancestors were to be found in the Carboniferous among the reptiles of the coal swamps, as for example *Petrolacosaurus*. This

was a small lizard-like reptile feeding on insects and other invertebrates. In the vertebrate group, these reptiles were close to the base of the food web. During the succeeding Permian period the paramammals, which had been semi-aquatic reptiles, became established as the dominant reptiles on land. Meanwhile the descendants of *Petrolacosaurus* changed very little and retained their place as the main vertebrate insect-eaters.

Two evolutionary lines, clearly distinct by the beginning of the Triassic, arose from these small generalized reptiles of the *Petrolacosaurus* type. The first, the lizards, held onto the insect-eating niche, although during the Middle Triassic several specialized forms evolved which colonized shorelines, first as scavengers feeding on arthropods, and thereafter venturing into the sea as fish-eaters. The second evolutionary

line became the archosaurs which took up the semi-aquatic niche that had been abandoned by the paramammals. At the beginning of the Triassic period, 225 million years ago, the first archosaurs (the group which gave rise to crocodiles, dinosaurs and birds) invaded the unoccupied semi-aquatic niche, feeding on fish and acting as predators, in much the same way as crocodiles, living archosaurs, still do. In other words, a branch of the reptiles returned to a semi-aquatic life, to which they had to re-adapt.

Aquatic or semi-aquatic animals need an efficient means of moving through water. Effective propulsion is achieved by undulating movements of the body, especially the tail. One of the trends that characterized the early archosaurs was an increase in the tail muscles: the tail became more and more powerful and more flattened on each side, until it became the major

▶ A reconstruction of *Cheirotherium*, which is based on data derived entirely from the footprints, because no other remains of the animal have yet been found. The restoration has been confirmed as accurate by the discovery of a complete skeleton of a very similar animal, *Ticinosuchus*.

▲ Prints from the fore- and hind-feet of the Triassic reptile, *Cheirotherium*, a possible ancestor of the dinosaurs. The footprints

shows very fine details of the texture of the skin on the soles of the feet. A number of lizard-like tracks and what appear to be

gastropod trails can also be seen on the specimen.

swimming organ, just as it is in living crocodiles. By itself, however, it was not enough. What the reptile needed was a means of providing immediate acceleration in water, after which the tail action could maintain the momentum. This thrust comes most effectively from the hind-limbs and, during the evolution of the archosaurs, there was a gradual increase in the length and strength of the hind-limbs as compared with the fore-limbs.

In the sprawling gait of primitive reptiles, such as lizards, the upper limbs project sideways and the main movements of the legs are from the knee so that only part of the limb is used directly for propulsion. The main purpose of the sideways-projecting upper limb is to form a kind of stable sling for the body. In the case of large semi-aquatic predators like the archosaurs, which preyed on animals that came down to the water's edge to drink, an initial burst of speed was essential. The more of the limb that could be used to this effect the better. The limbs were straightened, and the limb musculature was better utilized for providing thrust, just as in swimming where using only the forearm is less effective than using the entire arm from the shoulder. The result of the development of straighter limbs was that when the archosaurs ventured onto land, they were capable of walking with their bodies held completely off the ground, and with the hind-limbs markedly longer and stronger than the fore-limbs. The crocodiles illustrate this particular stage with their 'high walk', where the limbs are more or less straight and the body clear of the ground.

These developments, which took place during the Triassic period, were to have dramatic consequences. As these specializations for an active semi-aquatic predator continued, a stage was reached when the potential speed of the long hind-limbs on land was severely curtailed by the much shorter stride of the fore-limbs. Several living lizards, like the Australian frilled lizard, lift the forepart of the body and run on their hind-limbs alone when maximum speed is required. So too did the more advanced archosaurs; with the fore-limbs out of the way and with their long hind-limbs, impressive bursts of speed on land became possible. Furthermore, the heavy muscular swimming tail acted as a counterbalance to the head and trunk, so that the whole body was pivoted over the hip-girdle. These reptiles' adaptations to a semi-aquatic life had fortuitously pre-adapted them to movement on land as two-legged predators.

The first archosaurs to achieve this method of locomotion, as evidenced from the details of their limb-bones and -girdles, are known as 'dinosaurs'. They were exclusively meat-eating, as they had been living a crocodile way of life, and were introduced into a fairly stable modern-type ecosystem. The result of the appearance of these active carnivores was essentially a collapse of the established food web. They preyed on the majority of the paramammals which became extinct, with the exception of a few large heavily-built herbivorous paramammals, like the Dicynodonts, and some small ones, like the water-vole *Oligokyphus*, managed to continue for several million years. The really successful survivors were those that became progressively smaller and continued to feed on insects. These were the first mammals which, strangely enough, filled the same insect-eating niche of the distant ancestors of the archosaurs, such as *Petrolacosaurus*.

▼ *Proterosuchus*, a semi-aquatic feeder, was one of the first archosaurs.

▶ *Coelophysis*, a small lightly-built dinosaur, was the ancestor of a flesh-eating group of dinosaurs, the coelurosaurs.

▲ The crocodile 'high walk' with the limbs held more or less straight, unlike the normal sprawling reptilian gait.

THE AGE OF DINOSAURS

**Dinosaurs flourished in a large variety of shapes and sizes for 140 million years.
Most of these highly successful animals were herbivorous.**

The dinosaurs lived in a tropical and sub-tropical climate from the late Triassic to the late Cretaceous. These warm humid conditions were world-wide and dinosaurs seem to have inhabited every region on Earth, although no dinosaur remains have yet been found in the Antarctic.

The dinosaurs comprised two different groups, the Saurischia or 'lizard-hipped' reptiles and the Ornithischia or 'bird-hipped' reptiles. The hip-bones of the saurischians resemble those of many reptiles while the ornithischians are quite different.

The first dinosaurs appeared in the late Triassic when there was a preponderance of saurischians. The earliest saurischians were 1–2m long and exclusively flesh-eating. There were two groups, the more heavily-built carnosaurs and the lightly-built coelurosaurs. From a probable coelurosaur stock there arose the prosauropods, which increased in size to reach 12m in length. The early prosauropods had sharp serrated blade-like teeth for meat-eating, but before long, the teeth became peg-like, losing their serrations. This was because these dinosaurs had become plant-eaters which led to the largest ever animals, the brontosaurs or sauropods, some of which have been calculated to have weighed over 100 tonnes.

The carnivorous saurischians developed in two different directions. Some became heavily-built dinosaurs like *Tyrannosaurus*, others remained light but evolved different feeding strategies. The smallest and lightest forms such as *Compsognathus*, some 300 mm high, hunted lizards, as proved by the partially articulated lizard skeletons which have been found in their stomachs. They also probably scavenged at the kill of other dinosaurs. From footprints it seems that all the carnivores moved about in small groups.

During the latter part of the Cretaceous some 100 million years ago, the carnivores developed new types of feeding strategies. The most successful carnivore was the lightly-built *Deinonychus* or 'terrible claw', which possessed an enormous sickle-shaped claw, probably to disembowel its victims.

Some further evidence that confirmed this came from examining casts of the inside of the braincase of this group. The ratio of brain weight or volume to the weight of the complete animal surprised scientists as it was much greater than that of any reptile or other dinosaur. This was mainly due to the enlargement of the cerebellum, the part of the brain concerned with muscular co-ordination, which indicates that these dinosaurs had a highly developed sense of balance.

Another group of lightly-built lizard-hipped dinosaurs developed during the Cretaceous: these were the

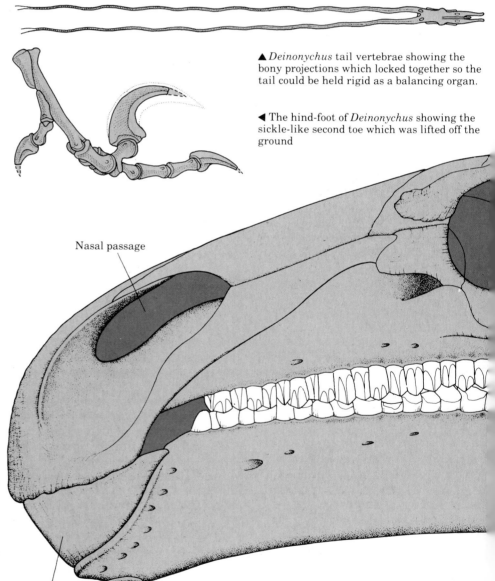

▲ *Deinonychus* tail vertebrae showing the bony projections which locked together so the tail could be held rigid as a balancing organ.

◄ The hind-foot of *Deinonychus* showing the sickle-like second toe which was lifted off the ground

Nasal passage

Predentary bone

▲ The skeleton of *Shangtungosaurus*, the largest duck-billed dinosaur known, on display in the Natural History Museum, Peking, China.

▲ Side view of the skull of *Iguanodon*, an advanced herbivorous dinosaur. This shows the greatly expanded opening of the nasal cavities connected with cooling, the ridged teeth and the dished-in side of the face, which indicates the presence of a muscular cheek. The predentary bone is the trademark of the ornithischians.

ostrich dinosaurs, such as *Ornithomimus*, which had lost their teeth but retained long thin fingers which could deftly snatch and grasp prey. As with the sickle-clawed dinosaurs, they were noted for their comparatively large brains and, in some cases, enormous eyes. These large eyes suggest that they may have hunted in fading light; presumably the prey was small furry mammals which came out to feed as night fell.

Most herbivores were ornithischians. They had as their distinguishing feature a small single bone at the tip of the lower jaw called the predentary. This bone is unique to this group of reptiles and was covered with a horny beak used for plucking and tearing plants. Footprints show that the herbivores lived in herds of up to 30 individuals. It seems that when they were on the move, the largest

animals would be positioned on the perimeter of the herd.

The first ornithischians, found in Upper Triassic rocks, were small (about a metre long), lightly-built bipedal forms, such as *Lesothosaurus* and *Heterodontosaurus*, known as the ornithopods. They gave rise to a line of heavily-built forms, such as *Iguanodon*, which evolved later in the Cretaceous into the large duck-billed dinosaurs or hadrosaurs, which were among the most successful of all the plant-eaters. The other bipeds were the less numerous boneheads or pachycephalosaurs. Their skulls are thought to have been thick because they indulged in head-butting, rather like sheep and goats today.

Apart from the ornithopods, all other groups of ornithischians were quadrupedal. Among these were the horned dinosaurs, the ceratopsians, like

Triceratops which originated from the lightly-built ornithopods. Their horns and spikes are now generally interpreted as having developed through trials of strength to determine leadership of the herd. Other quadrupedal dinosaurs had heavy bony armour, which served as effective protection against predators. There were variations in the development of the armour. Perhaps the most dramatic-looking of these was *Stegosaurus*, the 'roofed' reptile.

All these herbivores developed structures which are unknown in other kinds of reptiles, but which had evolved quite independently in the mammals. For example, these dinosaurs evolved secondary palates so that food could be held in the mouth and prepared for digestion while the animal was still able to breathe. While it is true that crocodiles also have secondary palates, it should be stressed that these developed for a different reason – to enable the crocodiles to breathe under water. Several groups of dinosaurs, particularly the duck-billed dinosaurs, had batteries of teeth that effectively ground down tough plant materials such as pine needles and cones. Coupled with this, there were muscular cheek pouches, otherwise known only in mammals.

In 1979, in the Upper Cretaceous rocks of Montana, USA, dinosaur nurseries were found which comprised raised mounds 3m in diameter, with 2m wide nests hollowed out of the summit. These contained the remains of 11 baby hadrosaurs, each a metre long, all with worn teeth, which showed that they had not just hatched but had been eating for some time. Their presence in such a vulnerable situation implies parental protection. Living reptiles, like crocodiles and some snakes, similarly protect their young. New evidence like this led to a re-examination of the dinosaurs.

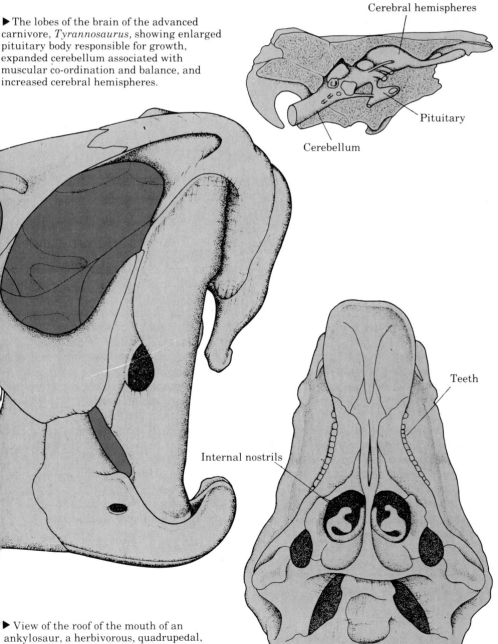

▶ The lobes of the brain of the advanced carnivore, *Tyrannosaurus,* showing enlarged pituitary body responsible for growth, expanded cerebellum associated with muscular co-ordination and balance, and increased cerebral hemispheres.

Cerebral hemispheres

Pituitary

Cerebellum

Internal nostrils

Teeth

▶ View of the roof of the mouth of an ankylosaur, a herbivorous, quadrupedal, armoured dinosaur. This shows the secondary palate and the internal nostrils set back behind the teeth.

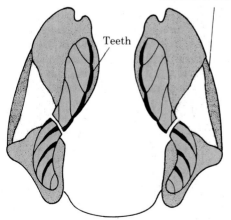

Cheek muscles

Teeth

▲ A cross-section of the jaw of a duck-billed dinosaur, or hadrosaur, showing the teeth and probable position of the muscular cheeks.

DINOSAUR SPEEDS

**Sometimes the only traces that a dinosaur
leaves are footprints. But these tracks can identify the
animal and tell us at what speed it was moving.**

The first serious reassessment in the 1970s of dinosaurs concentrated on a review of their limb structures. The proportions of the different parts of the limbs were calculated and compared with the limbs of living vertebrates, such as the ostrich, the elephant and the rhinoceros. The form and proportions of the limbs of the sauropods and the ceratopsians were similar to elephants and rhinoceroses. It therefore seemed reasonable to infer that their locomotion was comparable and that these dinosaurs would have been capable of moving at the same speed as an elephant, that is, up to 40 kph. Similarly, the limb proportions of the ostrich dinosaurs led to the conclusion that they were capable of speeds approaching 80 kph. These conclusions contrasted dramatically with the general view held of slow, sluggish, cold-blooded reptiles.

The question arises as to whether there is any means of determining the real speed of dinosaurs. While the limb proportions may be comparable to those of elephants and ostriches, did the dinosaurs have the appropriate musculature and metabolism to produce those speeds? There is direct evidence of the way in which dinosaurs moved in the footprints which they left. By examining them, the dinosaur can be identified. Points to look for are the number of toes, whether they had claws or blunt hooves, and whether they walked on all fours or were bipedal. From the size of the footprint and the relationship of the size of the feet to the complete skeleton, it is possible to work out the overall size of the animals that left their tracks. This evidence is crucial for determining the speeds of the dinosaurs.

The length of the dinosaur's strides can be easily measured and from the footprint alone it is possible to calculate the height above ground of the hip and shoulder. Maximum speed cannot necessarily be determined, but it is possible to work out the exact speed at which the dinosaur was moving when it left the footprints. A large animal moving fairly slowly will take long strides, whereas a small animal moving fast takes very small strides. What matters is not the actual length of the stride, but its relationship to the size of the animal concerned. This is called the 'relative stride' and is calculated by dividing the length of the stride by the height of the limb from the ground to its articulation with the limb-girdle. There is a relationship between the length of stride and body weight and speed; the formula applies equally to four-footed animals, such as horses and elephants, and bipedal forms, such as man and ostriches. This is shown in the accompanying chart where it is possible to read off the speed from the calculated relative stride. As the formula is the same for every animal, whatever its size, it

▲ Tracks at Queensland, Australia, showing what has been claimed to be the first evidence of a dinosaur stampede. There are footprints of small lightly-built dinosaurs which seem to be running at top speed and the tracks of a single, large dinosaur.

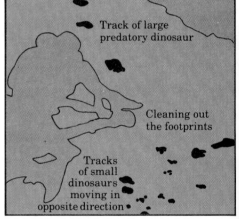

▲ Key to Australian dinosaur stampede showing the different footprints.

▲ Late Jurassic footprints at Swanage, Dorset, England, of two *Megalosaurus* dinosaurs.

applies to all dinosaurs.

A sandy beach is the best place to test this formula for yourself by walking and running; it can also be tested with a dog, or riding on horseback. The relative stride can be worked out and the speed deduced. This can be checked against the actual speed which has been timed with a stop-watch over a measured distance.

In one famous set of trackways of sauropods in the Paluxy River, Texas, USA, there are larger and smaller prints; the stride lengths are 2.5m and 1.6m, the height above ground of the hip-joints comes to 3m and 1.5m respectively and by dividing the stride length by the height, the relative strides come to 0.8 and 1.1 respectively. When read off the chart these suggest that the adult sauropods were walking at 3.6 kph and the young at 4 kph. This certainly gives an impression of a slow ponderous method of locomotion. It has been calculated from studies on the strength of bone that at speeds of 20 kph the forces exerted on the limb-bones would result in their fracturing. All this suggests that the giant sauropods could not have attained the speeds that can be achieved by today's elephants, no doubt because they were 15 times larger.

Tracks made by the large bipedal carnivore *Megalosaurus* from Swanage in Dorset, England, have a stride of 1.5m and hip height of 1.1m, giving a relative stride of 1.3 and indicating a speed of 4.3 kph, again a very slow rate of moving. Associated with the sauropods' tracks from the Paluxy River, Texas, is a trackway of a bipedal carnivore with a stride of 3m and hip height of 2m, giving a relative stride of 1.5m and a speed of about 8 kph. This individual was moving at almost twice the normal speed of a heavily built flesh-eater, suggesting it was approaching its maximum speed as it pursued a sauropod. Although the lighter carnivores moved at twice the speed of their heavily-built relatives, which made them more effective hunters, 8 kph is a very long way from the maximum speed of 80 kph that was postulated on the basis of the proportions of the limb-bones. This is now known to have been inaccurate.

A discovery made in Australia and announced in 1979 has been described as a dinosaur stampede. It showed some 3,300 footprints belonging to about 130 individuals. There were two kinds of small lightly-built dinosaurs; roughly half were herbivores and half carnivores, and they had been wandering about near the edge of a lake. There is also a single trackway of a giant flesh-eater with a stride of 3.7m and hip height of 2.6m which suggests a speed of 8 kph. The small dinosaurs appear to have panicked and rushed away from the water's edge, presumably to avoid being trapped by the large predator. Instead of placing their toes flat on the ground, they seemed to have been moving on the tips of their toes. The small carnivores had a relative stride of 2.4 and the small herbivores 2.5. The significance of these relative strides is that they are above 2, which means the dinosaurs were running. This is the first evidence that dinosaurs were capable of running, not just walking. The speeds come to 13 kph and 15 kph respectively and in view of the circumstances, it seems highly probable that these particular dinosaurs were achieving or coming close to their maximum speeds. The idea that the lightly-built carnivores could have reached speeds of 80 kph does appear doubtful. However in the last week of 1981 trackways of bipedal dinosaurs – carnivorous theropods – from Texas, USA, were described. These footprints showed that the animals were capable of a speed of 40 kph.

▲Dinosaur tracks at Peace River Canyon, Canada, photographed from a helicopter.

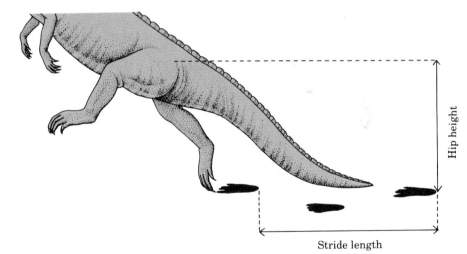

Hip height

Stride length

▲Sauropod and carnivore tracks at the Paluxy River, Texas, USA, following exactly the same course. The carnivore seems to be moving faster than usual – perhaps in pursuit.

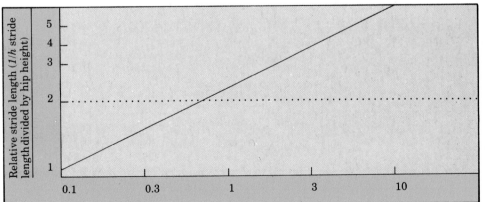

Relative stride length (l/h stride length divided by hip height)

▲Chart to calculate animals' speed. Velocity (speed) can be calculated directly from the relative stride using the formula $l/h = 2.3 (u^2/gh)^{0.3}$ (l = length of stride, h = hip height, u = speed, g = acceleration of free fall). Square of speed is proportional to linear dimension = u^2/gh.

HOT- OR COLD-BLOODED DINOSAURS?

Recently, scientists have been trying to prove that dinosaurs were hot-blooded as an explanation of the dinosaurs' success as a species.

There is no question about the dinosaurs having been exceedingly successful, but why should this have been so? Most scientists have believed that dinosaurs were cold-blooded reptiles, but a new theory was first proposed in 1968, by Bob Bakker of Harvard University, USA, that dinosaurs were 'endotherms', that is, hot-blooded animals whose energy was generated from within themselves, like birds and mammals. As vertebrates can only be active when warm, endothermy is regarded as an advantage. If an animal produces its own heat it can withstand cold, hunt during the night and be active over a long period. Reptiles can be at a disadvantage because they are cold-blooded and therefore rely on outside factors like sunlight to maintain a constant body temperature, which is why they are called 'ectotherms'. It follows that if the evidence for endothermy in dinosaurs were to be upheld, then it would no longer be possible to classify dinosaurs as reptiles.

The notion of hot-blooded dinosaurs seized both the popular and scientific imagination, attracting a lot of attention from the media on both

sides of the Atlantic from 1975 to 1980. There is circumstantial evidence for hot-bloodedness in dinosaurs; when considered separately, each set of data is not overwhelming, but in conjunction with one another, they can present a strong case. There are five main lines of argument for endothermy, but the real key to the question lies in the dinosaurs' great size.

The first argument relates to the stance and gait of the dinosaurs. Reptiles have a sprawling gait and happen to be ectotherms, whereas birds and mammals have the limbs held beneath the body in what is called the 'parasagittal' gait, and are endotherms. Dinosaurs have the same parasagittal stance as mammals, which could indicate that they might have possessed a similar metabolic system. However, there is no evidence that links this posture with endothermy. We know the dinosaurs' stance arose from the early archosaurs' adaptation to a semi-aquatic mode of life (see pp 128–9), and that this posture is advantageous as it uses less energy for both remaining stationary and moving. Furthermore, the limbs of a land animal of an appreciable size cannot

support its weight, unless they are positioned beneath the body. So it is clear that the reasons for dinosaurs assuming this posture may not have anything to do with being hot-blooded.

But in studying stance and gait, it was noticed that the overall proportions of the dinosaurs' limb-bones were similar to those of living mammals and birds. This led to the reasonable suggestion that their speeds might have been comparable. If this were so, dinosaurs must have had a high metabolic rate (the rate at which food is burnt up to produce energy) and were endotherms. It was asked: how could ostrich dinosaurs, for instance, have sustained a speed of 80 kph unless they were hot-blooded? Later, when dinosaur speeds were analysed, the evidence suggested that ostrich dinosaurs rarely exceeded 15 kph. There is evidence that some dinosaurs reached a speed of 40 kph (see pp. 132–3).

A recent argument in support of the notion of hot-bloodedness was based on the proportion of predator to prey. A hot-blooded animal needs more food to fuel its energy requirements than a cold-blooded one. In a modern-day fauna, an endothermic community

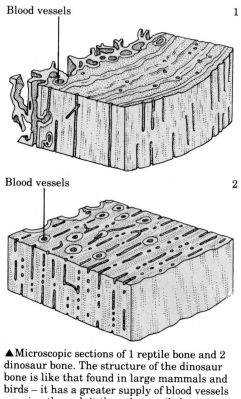

▲Microscopic sections of 1 reptile bone and 2 dinosaur bone. The structure of the dinosaur bone is like that found in large mammals and birds – it has a greater supply of blood vessels running through it than the reptile bone. This was taken as evidence that the dinosaurs were warm-blooded, but it is more likely to suggest that the dinosaurs had a rapid rate of growth.

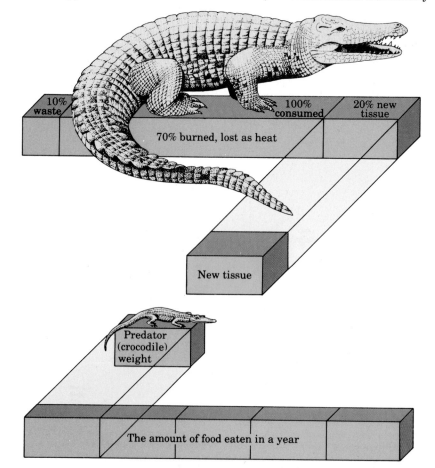

consists of 3% predators supported by 97% prey. In an ectothermic community the energy requirements are so drastically reduced that some 33% of predators can be adequately supported by about 66% of herbivores. It was claimed that from the fossil record the dinosaur faunas exhibited a predator-prey ratio that was similar to the mammalian or endothermic pattern and quite distinct from the typical reptilian ectothermic situation. At first glance this looks like a very convincing argument, but the evidence has to be examined closely. Not every animal alive at a particular point in time was preserved in the fossil record, so it cannot truly indicate the proportion of predators to prey. When it comes to the earliest geological records of dinosaurs, they happen to be of footprints which belonged to exclusively carnivorous forms. Again when the Cretaceous Mongolian discoveries (see pp 64–5) are examined, there seems to be an over-representation of carnivores. To date there has only been a single scientific study aimed at resolving this problem. A detailed analysis was undertaken of the dinosaurs of a later Cretaceous deposit in Canada in which the details of the parkland environment were able to be reconstructed; by virtue of this it was possible to calculate the amount of primary production, that is, plant material available, and then from the fossil remains, the proportion of

herbivores and carnivores. If the ecosystem was analysed in terms of the energy requirements of endothermic animals, there turned out to be far too many predators and not enough plant production to support the ecosystem. However, working on the hypothesis that the dinosaurs were ectothermic, the proportion of predator to prey turned out to be correct.

Another factor which has been considered is the geographical distribution of dinosaurs. Remains have been found in Arctic latitudes and the question has been posed as to how dinosaurs could have lived in the Arctic if they were not hot-blooded. However we know from studies of oxygen isotopes in limestone deposits that the Arctic climate during the Age of Dinosaurs was comparable to that of southern Europe today. The most telling piece of data are the fossil plants which were flourishing in the Arctic: these were semi-tropical. The climate of the Earth was warmer and more equable than it is now.

An indirect piece of evidence pointing towards dinosaurs' endothermy comes in the form of the first bird, *Archaeopteryx*. In many respects the skeleton of *Archaeopteryx* in minor details, such as the wrist, is so close to small lightly-built carnivorous dinosaurs that it is considered that birds are direct descendants of these dinosaurs. Birds are endotherms. Small-sized

endothermic animals, because of their large surface area compared with their volume, will cool down and warm up very rapidly in response to any outside changes in temperature, and so require some kind of insulation. This is hair or fur in mammals, and feathers in birds. The possession of feathers in *Archaeopteryx* is very strong circumstantial evidence that they provided just such an insulation. But the fact that the lightly-built carnivorous dinosaurs probably gave rise to *Archaeopteryx*, does not mean that their metabolism was the same. The mammal ancestors, the paramammals, did not have a high metabolic rate, nor did they have a furry covering. The acceptance of a relationship between two types of animals cannot be used as evidence in this way.

The final argument in support of dinosaurs being endothermic comes from studying the microscopic structure of dinosaur bone, which is similar to that commonly found among large mammals and birds; in many ways ox bone, for example, is indistinguishable from sauropod bone. This type of bone contrasts dramatically with the bone of living reptiles, which seems to have a very poor blood supply, with relatively few blood vessels running through it. The inference that has been drawn from these observations is that the metabolic needs of dinosaurs, birds and mammals were comparable, and were

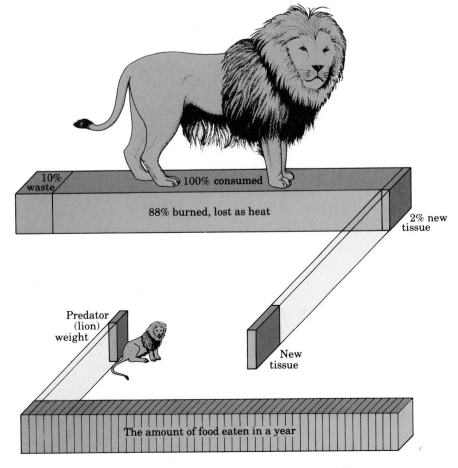

10% waste

100% consumed

88% burned, lost as heat

2% new tissue

Predator (lion) weight

New tissue

The amount of food eaten in a year

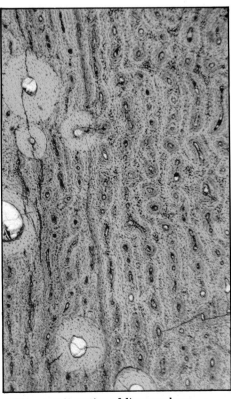

▲ Microscopic section of dinosaur bone.

◀ The metabolic rates of two predators, an ectothermic crocodile and an endothermic lion.

totally different from living reptiles. However, the large tank-like herbivorous paramammals, the dicynodonts of the Triassic which had a sprawling gait and were ectothermic, have exactly the same type of bone. Similarly crocodiles show the same kind of bone in parts of their skeleton. The structure of dinosaur bone, known as 'wire-netting' bone as that is what the network of blood vessels looks like, is found in large rapidly growing animals. This evidence suggests that dinosaurs had a rapid rate of growth compared with living reptiles. But as this bone is found in both hot- and cold-blooded animals, it is not possible to draw direct conclusions about their metabolic rate.

Whether an animal is hot- or cold-blooded, there is certainly an advantage in being able to maintain a constant internal temperature. This ensures that all the internal processes proceed at a steady rate and protects them from the effect of the fluctuations of temperature outside. The strategies used to achieve this vary. Living reptiles control their internal temperature behaviourally by basking in the sun to warm up and seeking shade or water to cool. Birds and mammals have various mechanisms both for losing and retaining heat, such as an insulating covering. The dinosaurs achieved the same end result by a quite different evolutionary strategy. They grew to a large size.

An increase in size effectively reduces the surface-to-volume ratio so that there is insufficient surface area to allow heat to be gained or lost readily. Assuming a daily temperature range between 22°C and 32°C, a small dinosaur with a cylindrical body a mere metre in diameter with a 5 cm layer of fat would have had an internal temperature fluctuating between 28.5°C and 29.6°C. The lowest temperature would have been at 09.30h and the highest at 19.30h. But the temperature at the surface of the skin would have been as low as 25.3°C at 05.00h, rising to 42.1°C at 13.00h.

In a warm equable climate, dinosaurs would have had all the advantages of being warm-blooded without having to have a high metabolic rate. Being an endotherm is very expensive as far as energy requirements are concerned. As living animals increase in size, their metabolic rate is reduced: an elephant has a very low metabolic rate compared with, say, a shrew. Indeed as

▶ The predator/prey ratio of a modern endothermic fauna.

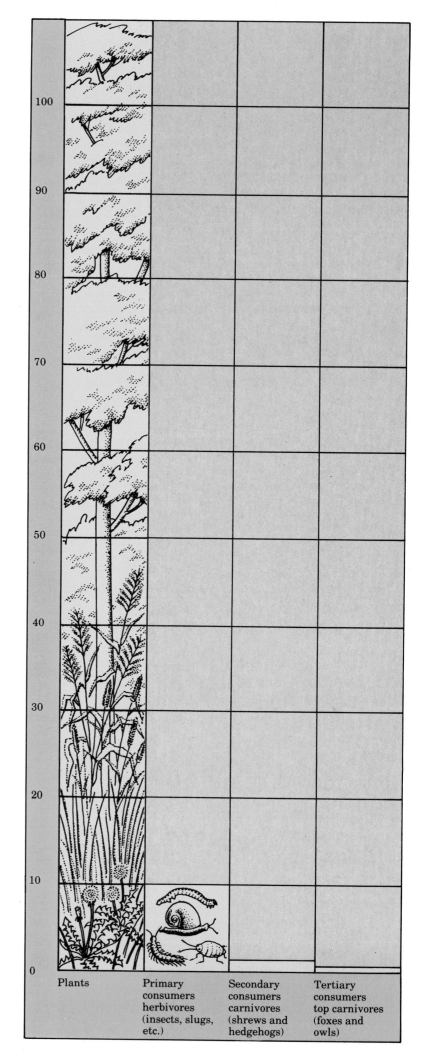

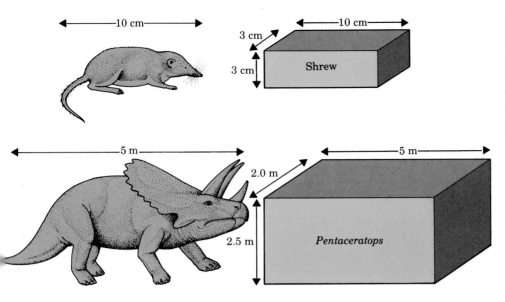

▲The surface area/volume ratios of the dinosaur *Pentaceratops* and an elephant shrew. The surface area to volume ratio of the shrew is nearly 80 times greater than that of the dinosaur.

▲To work out the ratios consider the body of each animal as a box.

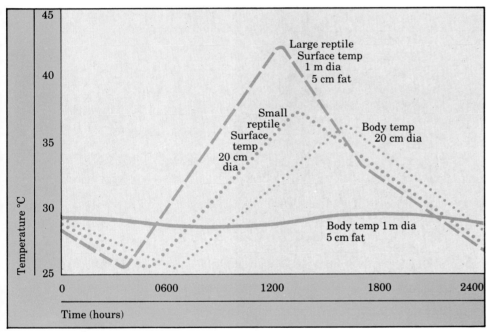

▲The diagram shows that unlike a small reptile, a large reptile's internal temperature remains nearly constant during 24 hours, although its surface skin temperature fluctuates considerably. The dinosaurs, by virtue of their size, were also able to maintain a constant internal temperature.

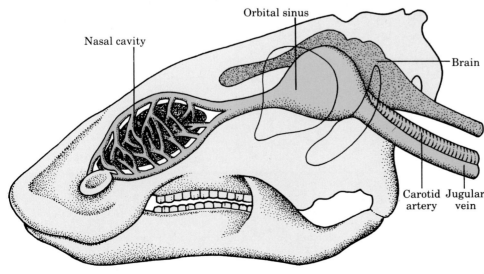

◀ Internal view of a hadrosaur (duck-billed dinosaur) skull showing large evaporative area in the nasal cavity and sinuses of venous blood which helped to regulate the brain's temperature.

animals grow, whether they began as endotherms or ectotherms, their metabolic rates converge so that, in the case of adult dinosaurs, they would inevitably have had a low metabolic rate. Endothermy in such large animals would have had disastrous consequences. First of all it would not have been possible to fuel a 100-tonne warm-blooded sauropod with sufficient food during 24 hours and for this reason alone endothermy in sauropods can be dismissed. An elephant has to eat most of the day to keep alive and one sauropod was equivalent in weight to 15 elephants. But the really serious problem for large animals is not producing heat but the very opposite.

The recent controversy over hot-blooded dinosaurs has been based on the wrong questions. The problem with dinosaurs was not creating heat, but rather how such large animals could dispense with the excess heat they produced simply by muscular activity. This is especially critical when it comes to the brain, because the brain is not situated deep in the body but close to the surface; a violently fluctuating temperature would damage the brain. In all the dinosaurs, there were large evaporative areas in the head region to help maintain a constant temperature. The nasal passages too, in some cases, were enormously complicated. Furthermore, there were large lakes or sinuses of venous blood close to the brain, as the blood acted as a heat sink, cooling the brain during the day and preventing the brain's temperature falling unduly at night (since the blood remained warm compared with the skin at night, and cool during the day). The very largest of the dinosaurs are characterized by long necks and tails, which served to increase the body surface area. The large triangular upright plates of *Stegosaurus*, the dorsal fins of *Spinosaurus* and *Ouranosaurus* were all temperature regulators.

It can be concluded from the above evidence that the dinosaurs had a constant internal temperature. They were homoiotherms, in other words they maintained the same internal temperature, but this was achieved passively, through the construction of their bodies, not through actively fuelling themselves. Dinosaurs were ectothermic homoiotherms. This would have been evident sooner, if scientists had begun by trying to answer a child's first question about dinosaurs: 'Why were they so big?'

FLYING REPTILES

Flying is a successful tactic for escaping from predators. Recent Soviet discoveries have cast light on how aerodynamic bodies must have evolved.

As well as conquering the land, the vertebrates managed to take to the air. However, true flight was only possible after the problem of maintaining a constant internal temperature had been resolved. A winged active flying vertebrate must be able to maintain a constant high internal temperature, otherwise in cold air all its metabolic processes will slow down and it will stall. The development of a flying body heavier than air is fraught with difficulties, so why should animals have taken to the air?

Flying started with the invertebrates. Insects first known from the Devonian had clearly taken to the air during the following Carboniferous, as evidenced by fossils of winged insects. One of the major habitats for invertebrate life besides the forest swamp floor was among the leaves of the forest canopy, which they must have originally reached by crawling. Flying probably developed as a means of avoiding the normal run of predators; and once in the tree-tops, the insects could enjoy an abundance of food. However, there is little direct evidence to suggest how insect flight developed, nor is there any kind of intermediate stage to show the possible route of this development. The fossil record of the insects is notoriously fragmentary, based only on a few remarkable insect-bearing levels, such as the Jurassic Lithographic Limestone of southern Germany.

One of the problems of living high up in trees is coming down to ground safely. A method of slowing down the descent was necessary. This can be achieved by extending the surface area of the body to create a type of parachuting effect. The living 'flying' squirrels, lizards, frogs, fish and even snakes, are all gliders. Either part of the body wall is extended, as in flying squirrels, or skin membrane is stretched over extended ribs, as in the flying lizards, to serve to increase their surface area. Vertebrates which are not active flapping fliers do not need to be hot-blooded.

The earliest aerial vertebrates were gliders. One of these was

discovered in 1978 in the Upper Permian rocks of Sunderland, England. It was part of a skeleton which had fine, greatly elongated ribs supporting an extensive gliding membrane. At the same time as this discovery, more complete examples of the same type of reptile were described from rocks of the same age in Madagascar, where the reptile was named *Daedalosaurus*. During the 1960s, a number of different gliding lizards, such as

Icarosaurus, were described from the late Triassic rocks of eastern North America and the Mendips region of south-west England. In these, the ribs supporting the gliding membrane were flattened and deep so that the membrane was probably stretched over both upper and lower surfaces of the ribs, like an aeroplane wing. It is evident that the rib-supported membranes evolved independently during three different geological

▲ These gannets show how birds, the descendants of *Archaeopteryx*, use their wings to soar.

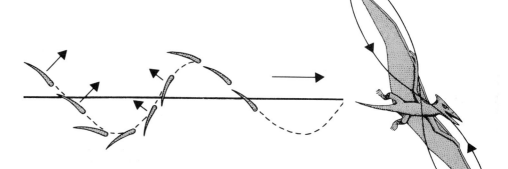

▲ The pterosaur flight pattern, which is not unlike that of the living fruit bat.

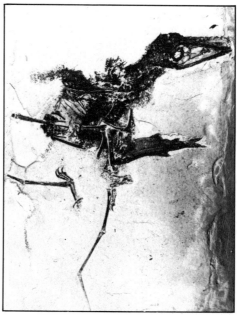

▲ Skeleton of the pterosaur *Sordes pilosus* showing preservation of the skin.

▲ A close-up of the downy or hairy covering of *Sordes pilosus*.

periods, as evidenced by *Daedalosaurus* from the Permian, *Icarosaurus* from the Triassic, and the living *Draco* (a gliding lizard from south-east Asia).

Another flying strategy was discovered in the reptile known as *Longisquama* or 'long scale' because of the elongated scales on its back. The fossil was recovered from Triassic rocks in Kirgizstan, Soviet Central Asia, by Dr G. Sharov of the Palaeontological Institute, Moscow, in 1970. Each scale was as long as the reptile's body and as the scales spread out sideways, they must have acted as a parachute. However, in evolutionary terms, the most interesting aspect of this specimen was that the scales along the trailing edge of the fore-limb were elongated to about seven times the normal scale length. It is thought that the feathers of birds could have evolved from such reptilian scales.

From the same Triassic deposits, Sharov described an even more astonishing reptile, which he named *Podopteryx* or 'foot-wing'. *Podopteryx* is significant because it is considered to represent a possible intermediate stage between the non-aerial lizard-like reptiles and the first actively flying reptiles or pterosaurs. Although it lived at the same time as *Icarosaurus*, it was a very different creature. *Icarosaurus*'s gliding membrane was supported by extended ribs, a structure which does not lend itself to detailed control and hence manoeuvrability. The preserved skin membrane of *Podopteryx* extended between the elbows and knees and from the hind-limbs to the tail to form a large rectangle. It is probable that such a type of gliding membrane between the fore- and hind-limbs was further extended to join onto the last finger of the hand, which is how the wing of the pterosaur is constructed. This type of wing membrane has more flexibility as it can be controlled by movements of the fingers and hind-limbs. Until *Podopteryx* was discovered, there was no way of explaining how the pterosaurs could have originated.

The year after this discovery, Sharov made another important find in the fine limestones of the Upper Jurassic in Kazakhstan, Soviet Central Asia. It was a pterosaur in which the skin covering was perfectly preserved. This specimen was called *Sordes pilosus*, meaning 'hairy devil', because it had a hairy or downy covering. The pterosaurs were active flapping fliers and this necessitated a high metabolic rate in conjunction with good insulation. It had been suggested as long ago as 1900 that, because of their flying strategy, they were endothermic and had an insulating covering. This logical inference was dramatically confirmed by the Soviet discovery.

Since pterosaurs are found in both marine and terrestrial deposits, they dominated the skies over land and sea throughout the Age of Dinosaurs. But towards the end of the Cretaceous, the descendants of *Archaeopteryx* (the first bird) had begun to occupy a number of important niches. These were algal-feeding flamingos, fish-eating cormorants and waders. The key to the success of the birds is that when they were grounded, they could tuck their wings into the side of the body and run around as bipeds. Pterosaurs, for all their initial success, had a wing membrane that was particularly vulnerable to damage; they were also quadrupeds so their wing membranes became entangled with their hind-limbs when they walked. This may be why they were superseded by a species better adapted to moving on the ground.

▲ *Sordes pilosus* and *Podopteryx* in life.

▲ The skeleton of *Podopteryx*, which may be an intermediate between reptiles and pterosaurs.

▲ The details of the wing membrane of the parachuting *Podopteryx*.

MARINE REPTILES

After millions of years on the land, some reptiles returned to the water in the Triassic. The two main groups of marine reptiles were the ichthyosaurs and the plesiosaurs.

During the Triassic period, the reptiles returned not only to freshwater but to the seas. The oceans were dominated by two contrasting types of swimming reptiles – the ichthyosaurs and the plesiosaurs. Fish-like ichthyosaurs were discovered at Lyme Regis by Mary Anning in 1811 (see pp 66–7), although Morton had described and illustrated individual bones as long ago as 1712 in his *Natural History of Northamptonshire*.

Numerous complete skeletons were discovered during the nineteenth century in England and Germany, all with an apparent break in the tail. This curious phenomenon was eventually understood when specimens from Württemberg in southern Germany were discovered by Bernard Hauff, a collector of fossils in the 1890s. He noted a black carbonaceous coating on the rock surrounding an ichthyosaur skeleton which he believed to be the skin – an idea dismissed as fanciful at the time. In 1892, however, Hauff found a complete *Ichthyosaurus* in which the entire outline of the skin

was preserved and the supposed break in the tail was seen to be where it naturally turned down, supporting a large fin above it. Along the middle of the back was a triangular dorsal fin. Discoveries of mothers with small unborn ichthyosaurs inside them, and even a mother giving birth to live young, showed that, unlike most reptiles, ichthyosaurs did not lay eggs.

Their diet was of fish and cephalopods. The cephalopods had hard tiny hooks on their tentacles which, when swallowed by an ichthyosaur, accumulated in its stomach; the fish scales, on the other hand, passed through the body and were incorporated in the fossil droppings or coprolites. It has been calculated from remains in the stomach that one ichthyosaur, less than 1.5m long, had eaten nearly 1,600 cephalopods.

In most popular books, ichthyosaurs are shown leaping out of the water rather like dolphins. The main skeleton was in the lower part of the tail, so the side-to-side movement of the tail not only drove the body

forward in undulating movements, but also pulled the tail downwards, which raised the forepart of the body. This meant that the main propulsive force from the tail would have driven the body up towards the surface and might have lifted it out of the water. The fore-limbs would have acted like the pectoral fins of sharks and helped regulate the position of the animal in the water.

The other group of marine reptiles were the plesiosaurs. The first

▲ A Jurassic marine crocodile skeleton from southern Germany.

▲ The complete skeleton of a nothosaur, a Triassic reptile; there are seven small nothosaurs surrounding the skeleton.

skeleton of a plesiosaur was also discovered by Mary Anning in Lyme Regis. These reptiles had exceedingly long necks and small heads with numerous long sharp teeth, which acted as fish traps. The tail seemed rather short, with a small dorsal fin, but there were four large paddle limbs for swimming. These creatures seem to have evolved from a group of Triassic reptiles known as nothosaurs which had webbed feet. These reptiles were fish-eaters and inhabited the coastal waters along the margins of the great ocean of Tethys.

The nothosaurs and plesiosaurs seemed to appear quite suddenly in the fossil record and there was no hint as to their ancestry until an important reptile was discovered in Madagascar

▲ Penguins use their wings in water, in the same action as was used by the plesiosaurs.

in 1981. This was the Permian reptile *Claudiosaurus germaini*, ancestor to the plesiosaur, which showed that the plesiosaurs evolved from the same basic stock that gave rise to the lizards and archosaurs.

From the primitive plesiosaurs, two major evolutionary lineages developed in the Jurassic. One was the pliosaurs, in which the neck became shorter and the head larger; the other, which continued to be known as the plesiosaurs, developed a very elongated neck. The large-headed pliosaurs had massive blunt teeth, the small-headed plesiosaurs had numerous fine teeth; this implies that their feeding habits were quite distinct. The difference in body shapes, too, suggests they were built for contrasting ways of life. We can ascertain what these were by studying the limb-bones and -girdles and the restored musculature of both types, which shows how they moved

In the pliosaurs, the muscles for pulling the limbs down and back were clearly massive, which suggests that this was the major swimming stroke; the muscles controlling the recovery stroke were particularly weak. To begin the swimming stroke, the paddle was held with the upper surface facing forwards at an angle, like a bird's wing in flight; when the limb was pulled downwards, it caused a power stroke

with an upwards and forwards thrust. At the end of the power stroke, as the water flowed over the limb, it rotated so that the blade presented the minimum resistance to the flow. The recovery stroke was completed by the dorsal muscles which pushed the limb forwards and then twisted the blade, ready for the next power stroke. The pliosaurs were powerful swimmers, probably capable of diving to depths of 300m to seek out the cephalopods on which they fed. Similar cephalopods live at such depths today.

In contrast, the long-necked plesiosaurs fed on fish and their movement in water reflected this. They had a limb musculature that was equally powerful for pushing limbs forwards and pulling them back. This meant that they were able to produce strong back strokes. The effect of a propulsive stroke with one limb and a back stroke with the other is to execute a very sharp turn. In fact the long-necked plesiosaurs seem to have been singularly ill-adapted for diving, but were well built for rapid twists and turns – a fact borne out by the detailed structure of the limb-bones. Their movements in water contrasted sharply with the sustained powerful swimming of the pliosaurs and would have allowed them to turn round quickly to snap up fish swimming by.

▼ Skeleton of a long-necked plesiosaur.

▲ The swimming pattern of the plesiosaurs.

▲ A restoration of the musculature of a short-necked, large-headed pliosaur, *Liopleurodon*.

▲ A remarkable ichthyosaur skeleton from southern Germany which shows the skin impressions outlining the large tail fin, large dorsal fin and the hydrofoil-shaped paddles.

THE EARLY MAMMALS

Early mammals survived during the Age of Dinosaurs by being nocturnal. When the dinosaurs became extinct, the mammals were able to take over their feeding niches.

The earliest example of a true mammal was found in the Jurassic Stonefield Slates at Woodstock, near Oxford, England, in 1764. However, no one realized what it was until the 1820s, and it was only named *Amphilestes broderipii* in 1871.

In 1812 two tiny jaws from the same site were brought to Oxford University's first Professor of Geology, William Buckland, who was convinced that they were from mammals. When their discovery was announced in 1824 it caused a sensation, because at the time it was believed that no mammals had existed during the Mesozoic. The mammals from which the jaws came were named later, one *Amphitherium prevostii*, the other *Phacostherium bucklandii*. Subsequently, fossil mammals have turned up in most parts of the world, including China, south Wales and southern Africa (Lesotho); and in 1981 discoveries were made in North America in Upper Triassic rocks.

The mammals seem to have survived during the Age of Dinosaurs by filling the nocturnal insect-eating niche, hitherto unoccupied. The evidence to show that they were insect-eaters lies in their dentition. The mammals' three-cusped molars were used to puncture and crush insects, after which the pieces were sheared into fine particles by the ridges joining the cusps. Evidence to show they were nocturnal lies in endocranial casts which describe the outer surface of the brain. It can be seen that the more developed parts of the brain were connected with smelling and hearing, the two senses most needed in a nocturnal life. An active, nocturnal animal would have to keep itself warm and endothermy was easier to achieve in a small animal than a large one; the small size of the first mammals suggests that they were both hot-blooded and nocturnal. The other nocturnal niche developed by the mammals was that occupied today by rodents. The insectivore and rodent ways of life continued through to the end of the Age of Dinosaurs. It is ironic that the time when the dinosaurs dominated the land coincided with the appearance of the animals which were to succeed them, the first mammals which never grew larger than rats for the whole 140 million years.

The mammals' antecedents were the mammal-like reptiles or paramammals, whose first representatives inhabited the Carboniferous coal swamps. When the reptiles colonized dry land, the paramammals became the most successful of them all and dominated the continents for some 70 million years, from the end of the Carboniferous to the mid-Triassic. The evolution of the paramammals into true mammals was very gradual and the boundary between the two could be drawn at a number of places. There were several features which characterized this transition and they did not all happen simultaneously. The main changes occurred in: the posture which affected the pelvis, neck and eventually brain; the jaws, teeth and development of the palate; the ear; the reproductive processes and suckling the young (the mammals derive their name from the Latin *mamma*, breast); and the achievement of endothermy, with the growth of a furry covering.

The earliest paramammals had a typical reptilian sprawling posture, but by the end of the Triassic period they had developed an upright gait, with

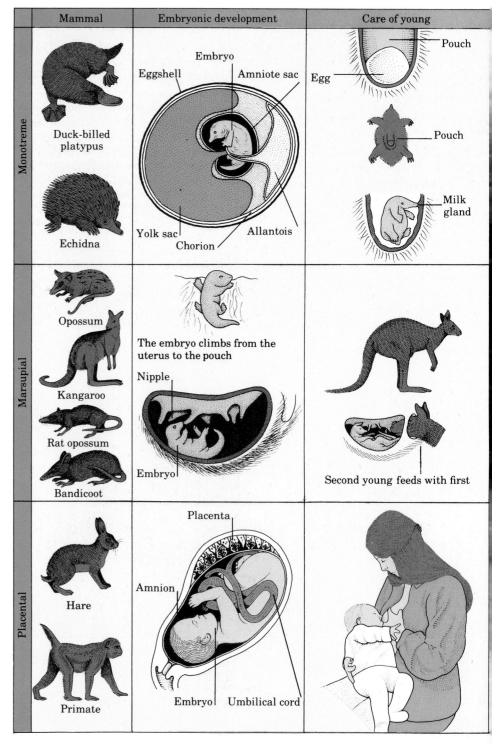

the limbs held vertically beneath the body. This series of changes is seen especially in the changing proportions of the bones of the pelvic girdle. Gradually the upper bone (the ilium) was developed at the expense of the posterior bones, which came to be directed backwards.

The effect of this improved locomotion was that the head was held far above the ground so that the mammalian neck developed. Furthermore, having the limbs held straight under the body meant that the stance and gait were more unstable than those of a sprawling reptile. The part of the brain concerned with muscular co-ordination and balance, the cerebellum, enlarged during this evolution to counteract this mechanical instability.

By far the most important changes took place in the jaws and teeth; it was these which led to the later mammal-like reptiles being classified as mammals. The first paramammals fed on fishes and amphibians in the Carboniferous swamps. Their jaw apparatus was essentially a fish-trap, merely preventing the prey from slipping out of the mouth. The tooth row was long and the teeth were numerous, small and sharp. With their emergence onto dry land, the main food of the larger individuals was other smaller reptiles. The entire process of acquiring food was different. The jaws and teeth had not only to hold the prey firmly as it struggled, but to dispatch it. An evolutionary sequence can be traced in which some of the teeth at the front of the jaw became longer to serve as stabbing weapons. The increased length of the crowns of the teeth necessitated a concomitant increase in the length of their roots. These were inserted in the upper jaw bone, the maxilla. Increasing the depth of this bone to accommodate the enlarged teeth resulted in changes in the proportions of the skull: the side of the face became longer, but the roof of the mouth remained where it was. As a consequence, when prey was gripped by the teeth, there was a space above the food through which air from the inner nostrils could reach the lungs. This space became the air tract, marked by a groove along the roof of the mouth; this groove was covered by a soft palate of skin which separated it from the alimentary passage. In some of the more advanced paramammals, this flap was invaded by body extensions from the inner part of the maxilla. This was the beginning of the secondary bony or hard palate that is a characteristic feature of all mammals, although it evolved independently in the ornithischian dinosaurs.

The most advanced paramammals had a complete bony floor to the nasal passage, separating the air and food passages. The floor of the secondary palate supported delicate scrolls of bone, the 'turbinals'; these were covered by mucous membranes which, in the living animals, both warmed and filtered the air. These same advanced paramammals also had, on the outer surface of the maxilla, a set of pits which seemed to mark the former position of whiskers. This provides circumstantial evidence that the advanced paramammals had an insulating covering of fur or hair and had achieved a degree of endothermy. Endothermy requires a means of losing and retaining heat to keep a constant body temperature. It also necessitates a secondary palate so that there can be a continual supply of oxygen to maintain a constant, high metabolic rate.

The ability to breathe and feed at the same time enables a young mammal to suckle. Detailed analysis of tooth replacement patterns among the advanced paramammals has revealed a striking reduction in the number of new teeth grown. Also teeth started to grow later than in the more primitive forms, so it is possible that this marks the beginning of suckling the young.

Another major change in the jaw led to the development of the

The duck-billed platypus and the echidna (spiny ant-eater) are monotremes, the most primitive living mammals. They are now found only in Australasia and New Guinea.

There is a view that these animals should not be classified as mammals, but as surviving mammal-like reptiles.

◀ The reproductive strategies of the three major groups of living mammals.

mammalian middle ear. When the history of the paramammals is traced, a gradual change in the shape of the lower jaw can be documented which is a direct reflection of changes in the arrangements of the jaw muscles.

The simple jaw-closing muscle of the primitive fish-eating paramammals came to divide into two major components, the one muscle pulling with a forwards and upwards motion, the other with a backwards and upwards motion. These two muscles working together gradually increased the force of the bite; the two muscle blocks, forming a controlled sling for the jaws, eventually made the bones of the jaw joint redundant, so that they atrophied. The bones at the back of the lower jaw were gradually reduced while there was an increase in the tooth-bearing bone at the front, the dentary. As this process continued, the dentary came into contact with the skull at the posterior cheek or temporal bone and a new jaw-joint connection came into existence. This change implies that there could have

been a point at which both the typical reptilian jaw joint and the typical mammalian jaw joint actually existed side by side. In 1958, Drs K. A. Kermack and A. W. Crompton, by a curious coincidence entirely independently described two unrelated paramammals which had achieved this state. Later some perfectly preserved paramammals from South America were described by Professor A. S. Romer, again with this particular intermediate stage perfectly preserved.

In order to follow the way in which the bones of the jaw could have become incorporated into the middle

ear, it is important to realize the relationship the bones have to one another. The sound-conducting bone of the paramammals, the stapes, was in direct contact with the bone of the upper jaw, the quadrate, on which the lower jaw articulated. The stapes was situated close to the jaw joint as was the eardrum. So it can be seen that in the advanced paramammals the eardrum and bones that were to become incorporated into the middle ear were already correctly positioned.

It is important to remember that by this stage of the evolution of the paramammals and mammals, at the

▶ Part of the skull of advanced paramammal *Thrinaxodon* from the Karroo, southern Africa, showing reptilian jaw joint.

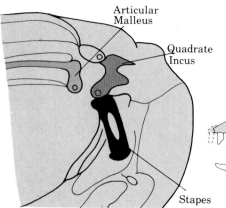

▲ Side view of the skull of the advanced paramammal *Probainognathus* showing both reptilian and mammalian jaw joints .

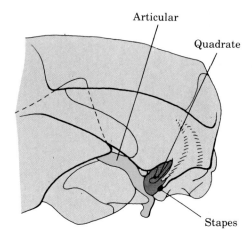

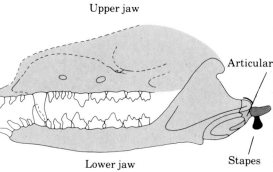

▲ Jaw of the early mammal *Morganucodon* from South Wales showing a reduced reptilian jaw joint existing with a mammalian jaw joint.

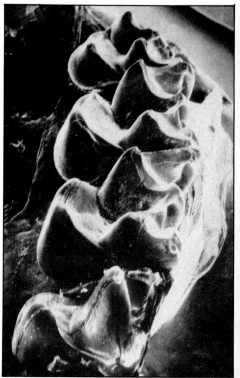

▲ Scanning electron microscope photograph of upper molar teeth from the primitive living primate *Tupaia* (tree shrew) which show the fundamental mammalian triangle of sharp pointed cusps.

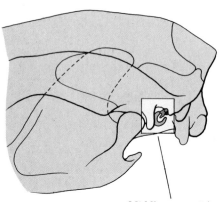

▲ Part of the skull of the primitive living marsupial American opossum *Didelphis* showing typical mammalian jaw joint and the

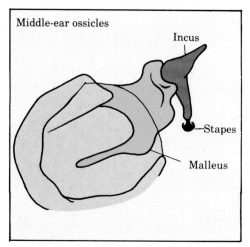

three middle-ear ossicles. A detail of the ossicles is shown.

▶ Late Cretaceous nocturnal mammals. Shown above is the North American *Purgatorius*, the first primate; the reconstruction of this animal is entirely based on the evidence of a molar tooth, which is above. Lower right is the herbivorous condylarth *Meniscotherium*.

end of the Triassic, these animals were very small. One of the features of developing small animals is that the sense organs such as eyes and ears are comparatively enormous. This means that during the early stages of development, there would have been a large eardrum and ear apparatus and, linked with this, the sound-conducting stapes in direct association with the quadrate of the upper jaw and the lower jaw hinge, the articular. Vibrations were carried from the articular (the mammalian malleus, the hammer), via the quadrate (the mammalian incus or anvil), to the stapes. The way in which the hammer and the anvil were attached to the skull dampened any sounds made by chewing or breathing, because these bones moved in line with the skull; this improved the detection of external sounds. It is difficult to imagine how the two reptilian jaw-joint bones could have avoided becoming incorporated into the middle ear, considering they were fortuitously connected to both the eardrum and the stapes.

One of the striking features of these first mammals is that there is fossil evidence to show that this transition from the reptilian to the mammalian ear occurred in a number of separate evolutionary lines of paramammals; this shows that mammals arose from several independent stocks.

The successful reproductive strategy of retaining the young within the mother and the development of the placenta to provide nourishment characterize the most advanced mammals. It is, however, important to remember that a placenta is found in the marsupials, and even in some fishes and reptiles.

The history of the paramammals documents the gradual achievement of mammalian status, but the exact stage at which one becomes the other is difficult to pinpoint. Classification depends on what features are considered the most important – suckling the young, the development of a secondary palate and cusped molar teeth, the changes in the jaws and ears, or the nature of the limbs and neck. Whatever feature is chosen, this will be merely an arbitrary line drawn by the scientists.

◄ The complete skeleton of a pantothere, a Mesozoic mammal, from Upper Jurassic rocks in Portugal. It is one of the most important mammal skeletons ever found; the bones in the pelvic girdle suggest that this animal was pouched just like the living marsupials. The elongated vertebrae of the long tail suggest that this was a tree-living animal which used its tail for balance as squirrels do.

THE DEATH OF THE DINOSAUR

**Many theories have been put forward to account for the disappearance
of the dinosaurs, but there is still no generally accepted explanation.**

By far the most dramatic event in the history of the vertebrates was the sudden extinction of the dinosaurs 64 million years ago. Numerous theories have been put forward; virtually any idea that one cares to propose will already have been suggested. Professor G. L. Jepsen's article, 'Riddles of the Terrible Lizards' (1964), quotes many widely differing theories, ranging from the plausible – deterioration in the climate or diet, disease, wars, anatomical or metabolic disorders, attacks by predators, evolutionary drift into senescent over-specialization, changes in the pressure or composition of the atmosphere, cosmic radiation, floods, drainage of swamp and lake environments – to the wild – development of psychotic suicidal factors, God's will, raids by little green hunters in flying saucers and lack of standing room in Noah's Ark.

For a theory to be taken seriously, it is not enough to consider the disappearance of the dinosaurs alone. Several other sets of data must be taken into account. For instance, the pterosaurs in the air and the plesiosaurs and ichthyosaurs in the oceans also died out at the same time. Moreover, so did the ammonites and some three-quarters of the microscopic organisms inhabiting the surface waters of the oceans. Such a wide range of extinctions may have been coincidental, but if a global explanation is being sought, it ought to account for these extinctions as well.

Even if a satisfactory theory can be constructed on all these counts, one still has to explain the survival of the mammals, birds, lizards, turtles, crocodiles, modern bony fishes, cuttlefish and their relatives, and modern land plants. These groups give no indication of anything untoward having happened during the disappearance of the dinosaurs. The problem is to explain not just the extinction of different organisms which inhabited a great variety of environments, but the remarkable selectivity of those extinctions. Can one find some event or process that affected these different forms of life?

Because the extinction of the dinosaurs remains a great mystery, an expert is always tempted to provide an explanation which derives from his particular field. The theories that attract the greatest publicity are naturally those which contain an element of drama. A recent theory invokes the impact of a meteor some 15 km in diameter striking the Earth. This is based on the discovery of a thin band of clay with a high content of iridium, which is exceedingly rare on the Earth; as the concentration is 20 times the normal, it has been postulated that this must have derived from an extra-terrestrial source, such as a meteor. The theory was put forward in 1976 by a physicist who

▲ Evidence for a change in the climate which could have caused the disappearance of the dinosaurs has been found in the rocks of Montana, USA. The restoration shows the moist sub-tropical forests in Montana prior to the extinction of the dinosaurs.

concluded that the debris from such an impact would have been thrown up into the upper atmosphere and blocked out the sunlight. The plants would have been unable to photosynthesize for several years and only small scavenging organisms feeding on the rotting animal and plant life would have managed to survive until the skies cleared. A variant of this theory, put forward in 1980, suggested instead a close encounter with a comet that would have heated the atmosphere; this would have created heat stress fatal for large land animals; there would also have been a rain of cyanide which would have destroyed most of the microscopic marine life.

In order to discover why the extinctions occurred at the end of the Cretaceous, the first task should be to examine what was happening at the time. Geologists speak casually of the sudden extinction of the dinosaurs, but when one is dealing with tens and hundreds of millions of years, 'sudden' can mean as long as 500,000 years. It tends to be forgotten that most of the animal and plant life that became extinct at the close of the Cretaceous had been in decline for the previous few million years, so that by the time of the final extinction, many groups were only represented by one or two kinds.

In 1981 a detailed geological study of the structures and chemistry of the sediments of the rocks preserved in Denmark that span the boundary between the Cretaceous and Tertiary, established that there had been a world-wide lowering of sea-level. The effect of this would certainly have been dramatic as far as the animals and plants of the shallow coastal regions of the continental shelves were concerned, particularly as there would have been a major extinction of the 'phytoplankton' (microscopic plants) which was at the base of the food web. The surface area of ocean waters would have been drastically reduced and less carbon dioxide would have been used as there were fewer plants to photosynthesize. Heat from the Sun would have reached the surface, but would not have been able to radiate back through the carbon-dioxide-enriched atmosphere. The temperature would have built up and large terrestrial animals would have been under severe stress. From isotope studies of carbon and oxygen from the calcium carbonate of limestones formed at this time, it can be shown that there was a sharp temperature rise. But after this, the overall global temperature fell progressively until the Pleistocene Ice Age, some 2.5 million years ago.

What happened to the land animals during this time? Studies in 1981 documented an initial gradual change-over of both the mammalian faunas and plant life. Subsequently, there was a tremendous reduction in the numbers of both individuals and species among the dinosaurs as small mammals increased in numbers. Detailed fossil collecting has established that the dinosaurs died out first in more northerly latitudes; they retreated southwards as the environment of tropical open forest was gradually replaced by temperate coniferous forests. As the mammals were endothermic, they could exist in colder temperatures; this probably accounts for their survival.

There is still no satisfactory theory to account for the extinction of the dinosaurs, but at least one can discount the extra-terrestrial catastrophic ones. It seems reasonable to interpret the disappearance of the dinosaurs as a process which occurred gradually over millions of years.

▲ The woodland of Montana had become temperate 10,000 years later when the dinosaurs became extinct and were replaced by a large fauna of small mammals.

THE RISE OF THE MAMMALS

The death of the dinosaurs left many feeding niches unoccupied. With little competition, the mammals gradually took these over and so became dominant.

The extinction of the dinosaurs meant that all the daytime feeding niches became available, but these were only gradually filled by the mammals. In 1981 studies of the rocks of the Montana Badlands, USA, revealed evidence as to which mammals became dominant in the Cretaceous, between the end of the Age of Dinosaurs and the beginning of the Age of Mammals. The mammals that typically occurred in the tropical forests at the same time as the dinosaurs were nocturnal. They comprised numerous types of pouched mammal, the marsupials, together with several species of the rodent-like gnawing mammals, the 'multituberculates' (so-called because their teeth had many cusps or tubercles). This fauna gradually reduced, mainly as a result of the dying out of the marsupials, which could not compete with the new influx of more advanced placental mammals. The only marsupial that survived was the opossum.

The Montana rocks show that, during the reduction of this mammalian fauna (which recovered slightly later) and the disappearance of the dinosaurs, more advanced mammals were coming in from other areas. These mammals are usually associated with the early part of the Age of Mammals; they included some multituberculates, several insect-eating mammals, a primate and some primitive clawed 'condylarths', which were the basic stock from which the later groups of herbivores originated. Although we know that their descendants became daytime feeders, we do not know when this occurred. As the end of the Cretaceous approached, this fauna gradually increased in importance and continued to increase well into the succeeding Tertiary era, with no sharp division.

The beginning of the Tertiary or Palaeocene period, 64–54 million years ago, witnessed the start of the great radiation of the mammals. Today there are 17 major divisions or orders of placental mammals, while 11 other major divisions from the Palaeocene have become extinct. At the beginning of the Tertiary, some elements of the fauna were cut off from one another on separate continents, so that for long periods of time, they evolved in complete isolation from each other (see pp 150–1). This meant that each continent tended to evolve its own unique mammalian fauna. At various times connections were established between formerly separated continents, which allowed faunal interchanges.

Essentially, on every continent certain basic types of mammal evolved, filling the major ecological niches of rodents, insectivores, herbivores and carnivores. Many quite unrelated animals from different evolutionary lineages were so similar that they could easily have been mistaken for one another.

One of the earliest mammalian niches, the rodent-like way of life, remained unchanged. The multituberculate rodents have survived until the present day which makes them the longest mammalian lineage so far, having lasted for 100 million years. The herbivorous condylarths

▲ A selection of the Eocene mammalian fauna of North America showing some of the ancestral types which gave rise to some of the dominant forms of later periods.
1 *Uintatherium* six-horned, sabre-toothed plant eater 2 *Palaeosyops* early titanothere
3 *Trogosus* gnawing-toothed mammal
4 *Hyrachyus* fleet-footed rhinoceros 5 *Mesonyx* hyena-like mammal 6 *Ischyrotomus* marmot-like rodent 7 *Stylinodon* gnawing-toothed mammal 8 *Helaletes* primitive tapir
9 *Sciuravus* squirrel-like rodent
10 *Smilodectes* lemur-like monkey
11 *Homacodon* even-toed hoofed mammal
12 *Helohyus* even-toed hoofed mammal
13 *Metacheiromys* armadillo-like edentate
14 *Sinopa* small archaic flesh-eater
15 *Orohippus* ancestral horse
16 *Machaeroides* sabre-toothed mammal
17 *Saniwa* monitor-like lizard 18 *Echmatemys* turtle 19 *Crocodilus* crocodile 20 *Hyopsodus* clawed, plant-eating mammal 21 *Patriofelis* large flesh-eater

evolved hoofed forms, except one line which developed into dog-like running carnivores, the mesonychids. The insect-eating niche also continued. It remained mainly nocturnal and some highly specialized forms, such as the ant-eaters, developed. The insect-eaters also gave rise to several groups of carnivore: there were small polecat-like forms as well as some larger cat-like types, the oxyaenids.

The earliest primates were also insectivores in the Cretaceous, but at the beginning of the Tertiary they occupied the niche of tree-living gnawers and nibblers, like squirrels. The primates were found in both Europe and North America and some primitive types must have reached both Africa and South America, where they became isolated and evolved quite independently. Most of the primates were rapidly ousted from their niche in the Eocene by the truly rodent primitive squirrels and died out first in Europe and later in North America. It is remarkable that the lineage which was eventually to lead to man almost became extinct very early on in its history, only succeeding as a result of the fortuitous isolation of some members on the continent of Africa. The early African rodents were ground-dwelling forms and not tree-livers, so they did not come into direct

competition with the early primates. (The South American primates evolved into South American monkeys, an evolutionary line distinct from the African monkeys and apes.) On the island of Madagascar the primitive primates were isolated and underwent their own separate evolution, filling a number of highly specialized niches. One of the most remarkable examples is the aye-aye, a rodent-like lemur, which filled the woodpecker niche.

In the northern continents the main plant-eating mammals of the Eocene comprised tree-living rodents and ground-dwelling animals of which rabbits and hares are characteristic. Before the latter group reached Europe from Asia, their niche had been filled in Europe by a miniature hare-like member of the camel family. The larger herbivores fell into three basic categories: medium-sized running types, slow, heavily-built types, and semi-aquatic forms. Within the rhinoceros family, for example, there were lightly-built running types, heavily-built forms which survive today, and a group called the amynodonts which filled the hippopotamus niche in North America.

In almost every geological age, there were giant herbivorous mammals in North America called the uintatheres, which were superseded by

the unrelated brontotheres. In Africa a giant herbivore with a double horn on its snout, the *Arsinotherium*, filled this role. One type of herbivore, known as the 'chalicothere', developed three enormous claws on its front and back feet. It was a member of the horse group, or odd-toed ungulates, and lived up until the Pleistocene Ice Age. Although it is known that the chalicothere was a plant-eater, the function of its claws is not yet known.

The flesh-eaters on all continents fell into a variety of basic types. There were the running hunters, such as packs of dogs; there were the cats which stalked and pounced on their prey; and there were the general scavengers, like hyaenas, which, although in the cat group, frequently ran down their prey. There were also some types of mongoose and polecat.

The main predators during the early part of the Tertiary were giant carnivorous flightless birds. The mammals, however, filled so many different niches that giant birds like *Diatryma* were severely restricted in their choice of habitat. The evolution of different carnivorous strategies by the mammals led to the birds being ousted from their niche. Nevertheless, similar albeit small predaceous birds, such as the marabou stork and secretary birds, still flourish on the plains of Africa.

▲ A selection of the Oligocene fauna of North America; the terrain was less wooded than in the Eocene. Among the mammals were early horses, early camels. The brontotheres were relatives of the horse and ancestors of the main ruminant lines; there were also some primitive pig-like forms.
1 *Trigonias* early rhinoceros 2 *Perchoerus* early peccary 3 *Mesohippus* three-toed horse

4 *Aepinacodon* remote relative of hippopotamus 5 *Hesperocyon* ancestral dog 6 *Protoceras* bizarre horned ruminant 7 *Archaeotherium* giant pig-like mammal 8 *Hyracodon* small fleet-footed rhinoceros 9 *Brontotherium* titanothere 10 *Subhyracodon* early rhinoceros 11 *Poëbrotherium* ancestral camel 12 *Protapirus* ancestral tapir 13 *Merycoidodon* sheep-like grazing mammal

14 *Glyptosaurus* extinct lizard 15 *Hyaenodon* archaic hyena-like mammal 16 *Hypsiodus* very small chevrotain-like ruminant 17 *Hoplophoneus* sabre-toothed cat 18 *Palaeolagus* primitive rabbit 19 *Ictops* small insect-eating mammal 20 *Hypsiodus* 21 *Leptomeryx* chevrotain-like ruminant 22 *Hypertragulus* chevrotain-like ruminant 23 *Ischyromys* squirrel-like rodent

THE MAMMALS FLOURISH

The beginning of the Tertiary saw a great radiation of the mammals.

The history of the mammals shows that they became the dominant form of life in the Tertiary. This situation has lasted until the present day, when the Earth is dominated by one species of mammal, Man. Primitive mammals were isolated on separate continents and evolved quite independently to fill all the available ecological niches. From a few unspecialized mammals, several independent evolutionary radiations took place. Even today the relatively unspecialized ones flourish.

By far the most successful of all the mammals in terms of numbers of species are the rodents and insectivores. There are 1,729 species of rodent and 374 species of insectivore. The flying mammals, the bats, are primarily insect-eating and include over 900 species. The other two successful mammal groups are the carnivores with 252 species, and the primates with 193, one of which is ourselves, *Homo sapiens*.

We know from Eocene fossil remains in North America that the major radiation of the large herbivorous mammals involved horses, tapirs and rhinoceroses. All three groups had both lightly- and heavily-built types. These odd-toed hoofed animals were dominant for tens of millions of years, until the mid Oligocene, whereas the even-toed ungulates were comparatively few in number. This was probably because they were small and unspecialized. During the Tertiary period, however the even-toed hoofed animals gradually diversified, giving rise to the pigs and their relatives, and eventually to the ruminants or cud-chewing mammals. The primitive ruminants included the camels; the advanced ones included animals such as deer, giraffes and the bovids (the sheep, goat, antelope and cattle group). Towards the end of the Tertiary, the even-toed ungulates underwent an almost explosive evolution and virtually replaced the odd-toed forms. Their success was due to the development of their digestive systems, and also to a bone, the 'astralagus', in the ankle which gave their back legs more spring when jumping. Of all the odd-toed ungulates, only the horse remains fairly successful today; the tapirs and rhinoceroses are close to extinction. Only 16 species of odd-toed ungulates still exist, compared with 194 of the even-toed group.

The success of the ruminants over the other large herbivores is probably due to their method of dealing with their food. After collecting their food, they retreat to places of safety to lie down and digest it at leisure. The food passes to the 'rumen', or first stomach, where bacteria and protozoa break down the cellulose of the plants. This material, the cud, is regurgitated and returned to the mouth, where it is chewed, and then passes to the real stomach. This method of dealing with plant material is highly efficient. The 'lagomorphs', animals such as rabbits and hares, have a different system: food passes through the stomach and when it reaches the caecum (a sac off the large intestine), the plant material is broken down by bacteria. However, there is no way in which the nutrients can be absorbed by the large intestine or colon; the food is thus voided and eaten a second time, after which the stomach is able to digest the material.

The mammals filled the same basic ecological niches in the different continents in strikingly dissimilar ways. Perhaps the most dramatic example among the living faunas is to be found in Australia. Until 40,000 years ago this continent was populated only by pouched and egg-laying animals. The marsupial 'mice' are actually insectivorous; there are

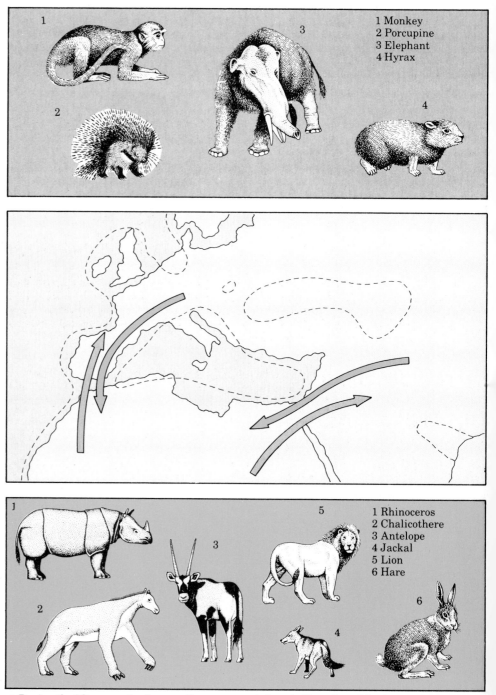

1 Monkey
2 Porcupine
3 Elephant
4 Hyrax

1 Rhinoceros
2 Chalicothere
3 Antelope
4 Jackal
5 Lion
6 Hare

▲ During the Miocene a land connection was re-established between Africa and Eurasia. This enabled some endemic African mammals to invade Eurasia and in the same way some Eurasian mammals moved south.

marsupial 'moles', cat-like and dog-like carnivores such as the Tasmanian devil and the Tasmanian wolf. The main herbivores are the kangaroos and wallabies which fill a variety of plant-eating niches and are unique to Australasia, with their jumping method of locomotion. No such type of large herbivore has developed on any other continent.

In South America, which was separated from North America at the beginning of the Tertiary, two unique forms of mammal evolved: the giant ground sloths and the massively armoured glyptodonts. Representatives of these two groups still survive in the armadillos and tree sloths. In addition to these unique types, South America

is notable for dramatic examples of 'convergent evolution'. This means that animals in one continent end up looking very similar to animals in the same feeding niches on other continents, although they are unrelated. Initially there was a mammalian fauna of possum-like marsupials, primitive unspecialized placental herbivores and primitive members of the armadillo family. From this basic tripartite stock the unique evolutionary radiation of South American mammals unfolded from the Palaeocene onwards. There were many South American marsupials which resembled other animals living elsewhere, which had the same feeding strategy. Good examples of this

convergent evolution were the marsupial sabre-tooth cat, *Thylacosmilus*; the marsupial *Groeberia*, resembling other rodent-like forms; the long-tailed leaping *Argyrolagus*, which looked like a jerboa; and the mole-like *Necrolestes*. The living insectivorous caenolestids of the Andes and the extinct marsupials all filled niches occupied on other continents by placental mammals. The most sensational examples of evolutionary convergence, however, were to be found among the South American Palaeocene ungulates. Instances are the proterotheres and macrauchenids, which so resembled horses and camels respectively that they would be indistinguishable from them at a first glance; the astrapotheres, which looked like the amynodonts (the amphibious rhinoceroses of North America); the beaver-like typotheres; the hopping or leaping hare-like hegetotheres; and the toxodonts which were rhineroceros-like, including both heavily-built and running forms. By Eocene times some 20 families and over 100 genera unique to South America were established, all of which are now extinct.

During the Oligocene, some 30 million years ago, primitive tree-living primates and ground-living rodents somehow reached South America; about 7 million years ago the coatamundi, a type of raccoon, arrived as well as field-mice. Since the Pliocene, a land connection between North and South America has allowed faunal interchange. The invaders from the north included the mastodons, horses, tapirs, peccaries, camels, deers, shrews, squirrels, mice, dogs, bears and cats. These mammals were very much more advanced than their southern counterparts, which could not compete with them and so gradually became extinct. However, some South American mammals, such as the giant ground sloths and the massive armoured glyptodonts, migrated northwards and survived until a few thousand years ago. They were accompanied by the most successful marsupial, the opossum; it is still moving up the east coast of North America, surviving on the refuse of man. The reconnection of North and South America ended the isolation of South American mammals and this evolutionary experiment.

A similar situation occurred in Africa but the re-establishment of a land connection between Africa and Eurasia took place at a critical juncture in the history of the Earth, during the Miocene 25 million years ago, when the grasslands spread.

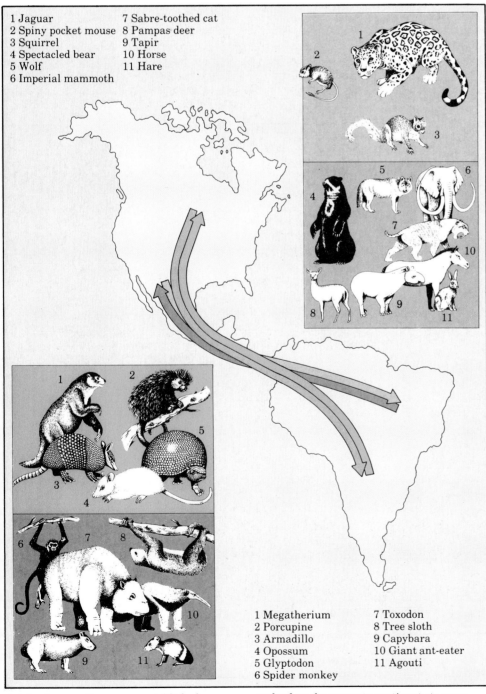

1 Jaguar
2 Spiny pocket mouse
3 Squirrel
4 Spectacled bear
5 Wolf
6 Imperial mammoth
7 Sabre-toothed cat
8 Pampas deer
9 Tapir
10 Horse
11 Hare

1 Megatherium
2 Porcupine
3 Armadillo
4 Opossum
5 Glyptodon
6 Spider monkey
7 Toxodon
8 Tree sloth
9 Capybara
10 Giant ant-eater
11 Agouti

▲ A land connection was re-established between North and South America during the late Pliocene allowing mammals that had evolved on the separate continents to move between them.

CRISIS FOR THE MAMMALS

The spread of the grasslands led to a great radiation of grazing herbivores. Simultaneously many carnivores and other types of herbivores died out.

The Miocene period, which began 25 million years ago, marked one of the major crises for mammals. It was not a sudden event, but rather one that developed slowly over millions of years, beginning in the Oligocene, about 35 millions years ago.

During the Oligocene, the average annual temperature in temperate latitudes was 18°C. During the Miocene it dropped to 14–16°C (today it is 9°C). The temperature of the world was changing considerably, apparently as a consequence of an ice-cap forming at the South Pole. A recent theory suggested that this was due to a tilt in the Earth's axis of rotation, but a view postulated in the 1970s is that the Antarctic continent actually moved from a temperate zone into a polar

region. One of the consequences of the climatic changes was the gradual reduction of woodland as the grasses spread. Although grasses had evolved at the beginning of the Tertiary, they formed a comparatively minor aspect of the plant life of the time. During the Oligocene, however, there was a marked increase in the amount of grassland territory, and by the Miocene there were vast tracts of grasslands. This development had a profound and double-edged effect on mammalian life.

When modern woodland plants are examined, it is evident that most of their energy is utilized in supporting themselves, with trunks and branches. In any tree or bush, the amount of material directly available for the provision of nutrients for animals is

comparatively small, comprising leaves and, at the appropriate time of the year, fruits or seeds. The woodland, therefore, is capable of supporting only a limited amount of animal life. With grasses, on the other hand, virtually no energy is used in supporting structures, so that almost the entire plant can be used for photosynthesis. Grasses have the added advantage of adaptability, as they can flourish in both drier and cooler climates, than woodland. When the climate is too dry or hot, the grass will die down, only to sprout up again immediately moisture is available – grass is very quick-growing. Virtually all the energy trapped by the plant can be converted to food, which is available to animals. The regions of the Earth which today provide most of the food

▲ During the Miocene the grasslands of North America were dominated by odd-toed ungulates such as horses and chalicotheres (*Moropus*). Primitive even-toed ungulates appeared including oreodonts, giant pig-like forms and early camels.
1 *Moropus* clawed mammal related to horses
2 *Daphaenodon* large wolf-like dog
3 *Promerycochoerus* pig-like oreodont
4 *Stenomylus* 5 *Parahippus* three-toed horse

6 *Steneofiber* burrowing beaver 7 *Syndoceras* antelope-like mammal 8 *Stenomylus* small camel 9 *Dinohyus* giant pig-like mammal
10 *Oxydactylus* long-legged camel
11 *Diceratherium* horned rhinoceros
12 *Steneofiber* 13 *Diceratherium* 14 *Merychyus* small even-toed hoofed mammal

for mankind are the grasslands, either directly in the form of cereals, or indirectly through grazing animals.

The development of grasslands meant that mammals had the opportunity to increase in both numbers and density. But in spite of this, major extinctions of groups of both plant-eating and flesh-eating mammals occurred. For animals that had relied on leaf protein for their sustenance, the tough siliceous grasses presented a problem because a diet of grass wears down the teeth. This would not matter unduly if mammals continually replaced their teeth as reptiles do. Unfortunately they have a severely limited number of teeth, comprising a first generation of incisors, canines, premolars and molars, and a second generation of incisors, canines and premolars. If these are all worn away, then the animal will starve to death. To solve this problem, mammals' teeth developed in two ways. The pattern of cusps became more elaborate so that, as the teeth wore down, a series of

enamel ridges were formed and their sharpness maintained. (This is because the harder enamel wears down more slowly than the other softer tooth tissue.) Also, the volume of the tooth was increased by the teeth being high-crowned and growing throughout life. This is seen in the open roots of the teeth of the advanced ungulates, such as horses and cows. With the spread of grasslands, grazers became the predominant herbivores and there was a major evolutionary radiation of the ruminants, especially the advanced bovines like the antelopes, cattle, sheep and goats.

Coping with grass as a food item, however, was only one aspect of the problem. Living in open grasslands meant lack of protective cover for grazers and a degree of conspicuousness not previously experienced — although this affected predator and prey alike. Successful grazers stand well off the ground so that they have a long and wide view of their surroundings. Furthermore, their

limbs are very long and their digits are reduced to either two or one. Not only can they observe potential predators from long distances, they are also sufficiently fleet of foot to make good their escape. Horns and antlers, although primarily concerned with display and contests, serve as effective weapons of defence, as indeed do hooves. During the Miocene limbs became specialized for speed; the lower bones lengthened while the upper limb-bones became comparatively short. The major muscles that moved the limbs became shorter and were concentrated near the shoulder and hip-girdles. In such an arrangement, a small contraction of the muscles causes the limb to move through a wide arc so that the foot at the end moves very rapidly with long strides. While a grazer's limbs are light and not powerful, they are capable of moving very quickly. It is this type of limb construction that allows antelopes, horses and cattle to escape from predators at great speed.

▲ By the Pliocene a radiation of even-toed ungulates related to deer and giraffes had taken place in the North American grasslands. The first grazing horses had evolved together with dogs and cats.
1 *Amebeledon* shovel-tusked mastodon
2 *Teleoceras* short-legged rhinoceros
3 *Cranioceras* cranial-horned even-toed hoofed mammal 4 *Synthetoceras* snout-horned even-toed hoofed mammal

5 *Merycodus* extinct pronghorn antelope
6 *Hypolagus* extinct rabbit 7 *Epigaulus* burrowing horned rodent 8 *Aphelops* long-legged rhinoceros 9 *Neohipparion* extinct three-toed horse 10 *Prosthennops* extinct peccary 11 *Pliohippus* ancestral one-toed horse 12 *Megatylopus* giant camel
13 *Procamelus* llama-like camel
14 *Hemicyon* bear-like dog 15 *Pseudaelurus* extinct cat 16 *Osteoborus* short-faced dog

AN ECOSYSTEM FROM THE PAST

The East African savannah, with its wide range of grazing animals, is a living example of the type of ecosystem which prevailed 25 million years ago.

The spread of the grasslands in the Miocene had a remarkable effect on the evolution of the herbivores. They became exceedingly varied and it is reasonable to assume that, even in open grasslands, they were not all competing for exactly the same food items. One grassland region that still supports a large variety of herbivorous animals is East Africa. The majority of grazers are even-toed ungulates, the zebra and rhinoceroses being the only survivors of the odd-toed ungulates. An enormous number of different kinds of antelope are found here.

A study of these living animals shows that they are all highly specialized in their food preferences. Grass is not a uniform material and the savannah grassland provides a whole variety of separate food items. The zebra feeds off the coarse heads of the shoots, the gnu and wildebeest feed off the leaves, and the gazelle and antelope feed off the seeds and shoots at ground level. There are also various browsers, from the giraffe to the gerenuk, feeding off different levels of trees and shrubs scattered throughout the grasslands. Thus there is a tremendous variety of animals living today in the equable climate of East Africa. This suggests that the wide variety of herbivores that evolved with the advent of grasslands were also likely to have been highly specialized feeders that did not compete unduly.

The plains of East Africa are a living example of the original type of ecosystem that came into being with the grasslands. They provide a model by which the extinct faunas of other parts of the world can be interpreted. In North America, for instance, there was a radiation of bovids, like *Hayoceros* and *Proantilocapra*, belonging to the pronghorn family of the higher ruminants; these are recognized by their strange and varied antlers and horns. Like the large number of grass-eaters in Africa, they probably occupied different niches on the grassy plains of North America.

The spread of grasslands in a reasonably equable climate led to a vast increase in the variety of herbivores. It also presented both carnivores and herbivores with new and somewhat similar problems. The major difficulty for the carnivores was that they could be observed from considerable distances. This led to the extinction of the major flesh-eating mammals that had been dominant during the early part of the Age of Mammals – animals such as the cat-like oxyaenids and the dog-like

mesonychids. Only one group of carnivores, the 'hyaenodonts', managed to survive for some time in the new situation because they were scavengers. There was simply no way in which a carnivore could walk up to its prey unseen as it could in forests. As a result, the carnivores had to rely on their intelligence to survive.

Apart from the small generalized carnivores similar to polecats and mongooses, the flesh-eaters have adopted two basic hunting strategies; both involve intelligence. The cats use stealth and cunning when hunting. They stalk their prey on padded feet and then pounce. The victim is either dispatched with a stab of the front canines or a bite. The sabre-toothed tiger used the first strategy, whereas the *Pseudaelurus* (similar to the modern leopard) used the second.

The hunting strategy developed by the dog employs teamwork. The prey is first isolated from its herd by a pack of dogs. It is then chased to and fro until it weakens, at which point the dogs close in and bring it down, tearing at its underbelly. Dogs are predators which hunt in groups and are therefore very much social animals. One of their key attributes is communication. Each pack has a leader and a hierarchy based on ability. Individual personalities tend to be highly varied, but during the chase they will be submerged for the sake of the common endeavour. Their pack instinct enables dogs to accept the leadership of a human being and be trained.

Cats and dogs are very adaptable, as well as highly intelligent; they owe their success to these two factors. They are not specialized in their food preferences and are noted opportunists. In this regard, they differ from the herbivores of the grasslands.

The beginning of the ice ages in the Pleistocene brought a change in the food available throughout the year. This meant that the herbivores in higher latitudes could not continue as highly specialized feeders. The bison migrated from Asia to North America via the Bering Land Bridge, and in a few tens of thousands of years replaced most of the native ungulate population; only the living pronghorn survives as a witness of the former radition of its group. The plains of North America came to be dominated by bison, to the virtual exclusion of all other ungulates, because the bison were unspecialized feeders. This was the pattern that evolved during the ice ages: success no longer lay with the specialists, but with the generalists. The modern world is dominated by fewer types of mammal, but those that flourish are highly adaptable and variable; this is the new evolutionary strategy that has marked the past two or three million years. Although one may regret the absence of the rich variety of species which existed in the past, one example still survives on the plains of East Africa. This allows us to glimpse what an ecosystem must have been like 25 million years ago.

▼ A dik-dik, one of the smallest antelopes of East Africa.

▲ Zebra and topi in Kenya, East Africa, part of an ecosystem unchanged since the Miocene.

◄ The East African grasslands provide food for a large variety of herbivorous mammals which are highly specialized in their feeding preferences. Shown here are dik-dik which feed close to the ground; the gerenuk stands on its hind-legs to reach leaves, while the giraffe feeds from the uppermost parts of the foliage; the Colobus monkey feeds from the leaves of the highest trees.

CHAPTER 5

HOW TO LOOK
AT FOSSIL MAN

Man belongs to a highly variable, adaptable species
that has come into its own over the past three
million years, especially during the Pleistocene ice ages.
Fossil remains of man have only come to light
over the past 130 years. When we compare ourselves
to these early specimens, we have evidence that man has
evolved; for instance, we are taller and have larger brains.
If we look at man's ancestry, we can see that his
physical characteristics are derived from an arboreal
life which were subsequently modified as a result
of his descent from the trees. Living as a member of
a hunting group on the plains encouraged
man to develop as a social being.

DISCOVERING THE ORIGIN OF MAN

Fossil evidence of early man comes from Europe, Asia and Africa. A link can be traced from the Miocene ape-like Ramapithecus to Homo.

Africa is always regarded as the birthplace of man, but this depends on how one defines man. However, we can be certain that the primates that evolved into apes and man did so in Africa during the early part of the Tertiary, when the continent was cut off from the rest of the world. Fossils of these early stages are not common, but some have been discovered in rocks of Oligocene age, some 35 millions years ago, from Egypt. *Aegyptopithecus* is the kind of animal from which the line leading to apes and men is likely to have derived.

It seems very probable that all these early primates were forest-dwellers, just as the South American primates have always been. With the spread of the grasslands, some lightly-built ape-like primates came to inhabit open savannah during the Miocene, 25 million years ago. It was at this time that Africa re-established its physical connections with Eurasia and these ape-like animals, known as the 'dryopithecines', spread to Europe and Asia. It was from these grassland-living dryopithecines that the line

which may have led to man emerged. Since dryopithecines are found in all three continents, it is impossible to say with certainty where mankind may have originated, but to date, most of the fossil evidence has been discovered in Africa.

Among the dryopithecines in Africa, Asia and Europe there is one called *Ramapithecus* which shows features such as reduced canines and the teeth arranged in a curve around the jaws, which are indicative of the line leading to man, although recent research suggests it may in fact be on the line leading to the orang-utan. By contrast, the other dryopithecines had the cheek teeth in two more or less parallel rows; these and other characteristics suggest that they were already on the line leading to modern apes, such as the chimpanzee and gorilla. *Ramapithecus* has been discovered in India, Pakistan, Turkey, Greece, Hungary and Africa.

The form that is considered the link between the dryopithecines and the genus *Homo*, man, is *Australopithecus*, meaning 'Southern

ape'. This was first described in 1925 by Prof. Raymond Dart on the basis of the skull and attached lower jaw and endocranial cast of the brain of a young individual. From this fossil, Dart postulated that the ancestor of man would have had a small brain and human-like jaws and would have walked upright. From his study of the brain cast he insisted that, although it had the volume of a chimpanzee's brain, the proportions of the parts were like those of the human brain. This interpretation was dismissed by the contemporary experts in anthropology and anatomy. At the time, it was firmly accepted that the evolution of

▲ The Taung skull, discovered in 1924, was the first known *Australopithecus* skull. It is the skull of a juvenile with a perfectly preserved endocranial cast and the complete dentition.

▲ The diagrams show Raymond Dart's interpretation of australopithecine tools and man involved the following steps – first the development of a large brain, then the change from an ape-like jaw to a human one, and finally the acquisition of a fully upright posture. This belief was based on a mistaken understanding of Neanderthal man, who had a human skull and jaw, but an ape-like, shambling gait; and also Piltdown man, discovered in 1900 was noted for his human skull and ape-like jaw.

It is interesting that an essay written by Friedrich Engels in 1876 (and published posthumously in 1925) had postulated that the upright posture and jaws would have developed before the brain increased in size. Engels also stressed the significance of the transition to a meat-eating diet whereby man became a hunter and made the first tools which were weapons. Dart, too, insisted that these were important factors in the

weapons. Clubs, daggers and scrapers were made from bones of hoofed animals.

transition from ape to man. Although most of the academic establishment did not take Dart seriously, the Professor of Geology at Oxford, W. J. Sollas, came out in his support. Subsequent discoveries vindicated Dart's interpretations.

Since that time numerous fossils of 'australopithecines' have turned up in Africa. By far the most sensational is not in the form of bones but footprints, 3.6 million years old. They were discovered by the anthropologist Mary Leakey (b. 1913) at Laetoli, Tanzania, in 1978. Two trails of footprints have been preserved in a mud of volcanic carbonate ash which set solid; they are thought to be of a man walking, and a female holding the hand of a child and stepping into the prints of the man walking just in front. Clearly this human family was walking upright.

Among the australopithecine remains in Tanzania are primitive stone tools fashioned from rocks that have come from sources tens of kilometres away. These early men made tools of bone and stone and even constructed simple circular shelters. Dart has provided convincing evidence that the australopithecines used the long bones of antelopes as weapons; they also crushed the skulls of baboons, and on occasions of one another, although this interpretation has been disputed. It has recently been

recognized that chimpanzees fashion tools, so the definition of man as 'the tool-maker' is no longer sufficient to distinguish him from the apes. The vital distinction is that the first men made tools for future use; they were not *ad hoc* constructions because there is evidence that they were transported from the places where they were made. The use to which the tools are put should also be considered: the tools made by the first men were for killing, so man can be defined as 'the weapon-maker'.

There seems little doubt that if man is defined as a weapon-maker, then the australopithecines come into the category of man. However, there is considerable controversy regarding the different types of australopithecine which have been described. There are both lightly-built and heavily-built types. Nevertheless, their cranial capacity appears similar and, although the size of the teeth and jaws differs considerably, it has been calculated that a similar force must have been exerted for the animal to bite.

It has now been established that modern man emerged from among the more lightly-built australopithecines, in particular the most advanced type now generally designated *Homo habilis*. He then progressed via the intermediate species *Homo erectus*, finally to become *Homo sapiens*.

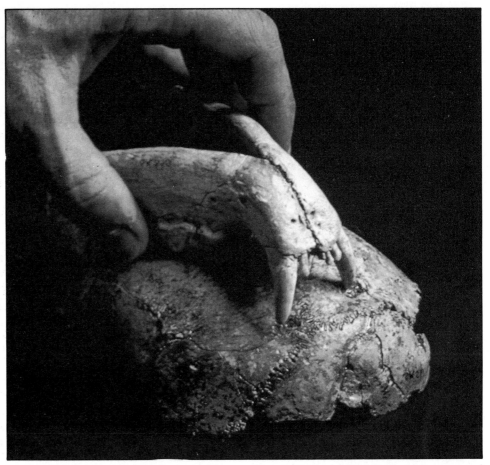

▲ A skull of another juvenile *Australopithecus* from Swartkrans, South Africa which has two dents in the back of the skull which correspond with the canine teeth from a leopard skull.

LOOKING AT OURSELVES

Modern man has several characteristic features which his ancestors evolved because of their life in the trees. These were modified as man became a hunter.

One way of investigating man's ancestry is to study the sequence of fossil men; the artifacts they left behind will provide evidence of their activities and behaviour. However, there is an alternative approach: we can look at ourselves to see if there is any surviving evidence of our ancestry in our anatomy.

For many centuries it was believed that man was a special creation and that his physical structure was part of a divine plan, created in the image of God. A closer examination of the structure of man reveals overwhelming evidence of faulty design. The most obvious error is that the viscera are supported by a strip of muscle that runs in the correct direction for an animal on all fours, but does not provide adequate support when held in an upright posture. Certain regions of the body, like the groin, take undue strain and the tissue may rupture in places, resulting in a hernia. This arrangement suggests that the features that characterize man have been superimposed on an existing pattern, one designed for moving and standing on all fours.

The most obvious characteristic of man is the rounded skull containing a large brain. There are other features which explain much of his success, and which would be difficult to understand if he were merely a special creation. For instance, compared with most mammals, man's sense of smell has been reduced enormously and is of much less importance. Because of this, the snout has become shorter and the nasal passages have taken an almost right-angled bend which encourages innumerable nasal ailments. The jaws have followed the snout in becoming shorter, leading to overcrowding of the teeth and a reduction in number from 44 in primitive mammals to 32 in man.

Other characteristic features of man are his eyes, which have rotated so that they face forwards. The fields of vision of both eyes overlap so that they can focus on the same object and see it in three dimensions. The value of this can be judged by trying to pick up objects quickly without moving the head and with one eye closed. The development of binocular vision is critical for judging nearby objects and their exact distances. This feature is shared by all the primates and is connected with living in trees. For a tree-dweller larger than a squirrel, the ability to judge distances accurately in order to leap from branch to branch is vitally important. This means that sight has become one of the major

sense, while the sense of smell has become less important, with a concommitant reduction in the snout. Muscular co-ordination goes hand in hand with sight; thus the cerebellum, the part of the brain concerned with muscular co-ordination and balance, has increased in size.

Man and other primates possess an opposable thumb, which enables the

▼ **Human features inherited from A our arboreal past B the descent from the trees** (1A) Binocular vision (2A) Reduced nose, shortened jaw and developed chin (3A) Collar-bone retained, arms can support weight (4A) Forearm rotates, opposable thumb, nails (replacing claws), sensitive finger-tips. (5B) Breasts. (6B) Pendant penis. (7B) Rounded buttocks. (8B) Non-opposable large toe. (9B) Feet flat to support weight (10B) Upright stance. (11B) Hair on head, facial hair on male, axial and pubic patterns.

▶ The relative proportions of tongue and larynx in *Homo sapiens* and *Homo erectus* suggest that modern speech was not possible for *Homo erectus*; he could probably only make a range of sounds comparable to those made by a human baby.

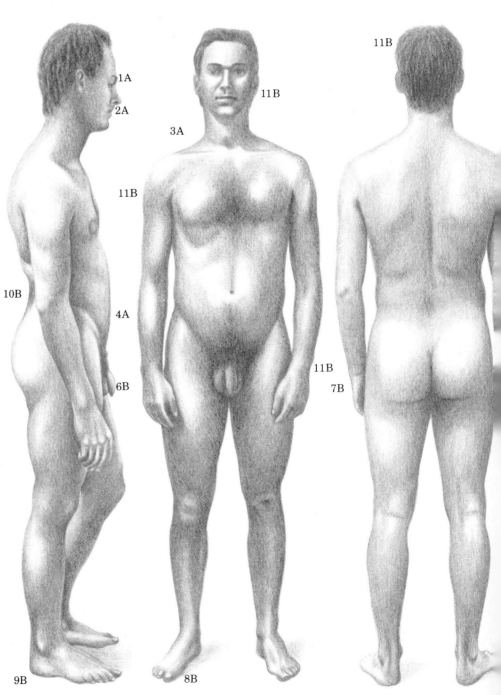

hand to grasp objects – thus to 'manipulate' the environment – a function usually performed by the teeth. The ends of the fingers are not armed by claws or hooves, but by flat nails so that the sensitive fingertips are able to perform delicate operations. The forearm contains two bones; the bone on the same side as the thumb, the radius, can rotate on the ulna and hence rotate the hand 180°. These features in present-day man can best be explained in terms of adaptations which allowed our ancestors to live in trees. In this environment the animals can leap among the branches, pick fruits and collect birds' eggs and fledgelings.

Besides these features, there are others which are not shared by the rest of the primates and which serve to distinguish man from his close relatives. In the higher primates, including man, the upper arm and shoulder region is sufficiently powerful to take the weight of the body. This points to the ability to swing from branch to branch, a mode of locomotion known as 'brachiation'. Once an animal habitually uses its arms for movement through the trees, this results in a differentiation between the roles of the hands and feet. However, when man descended from the trees, he landed on his feet. The large toe ceased to be opposable and the shape of the entire foot was adapted to support his weight. With the body being held in an upright vertical position, the articulation of the head on the vertebral column altered significantly. Instead of the back of the head being in line with the backbone, it became oriented at right angles to it. This made it impossible for the lower jaw to be lowered and the mouth opened without crushing the windpipe. If the fingers are placed on the face immediately in front of the ears and the mouth opened, it is possible to feel the jaw slide forwards out of its socket as it is lowered; while this prevents the windpipe from being crushed, it does make the jaw more vulnerable. Another distinguishing feature is that the genitalia of the male hang freely instead of being held in the body wall, a further consequence of man becoming a vertical biped, and again introducing a certain vulnerability.

The physical features of man indicate an initial arboreal existence, followed by a descent from the trees. This latter stage took place during the Miocene, about 25 million years ago. Other features of man are probably related to life in open grassland. For instance, the need to see over long distances when standing would have ensured that sight was still an important attribute. Other obvious distinguishing features of human beings include: the different patterns of facial and pubic hair; the turning outwards of the lips so that the skin just inside the mouth turns slightly onto the external surface of the face; and the formation of breasts on the upper region of the anterior surface of the body. These are all sexual recognition signals, and red lips in particular are supposedly indications of sexual receptivity. These signals are anteriorly placed because, when upright grassland primates sat down, the sexual skin would have been hidden from view. A consequence of this is that human beings generally mate face-to-face, which helps to cement the bonding of the individuals.

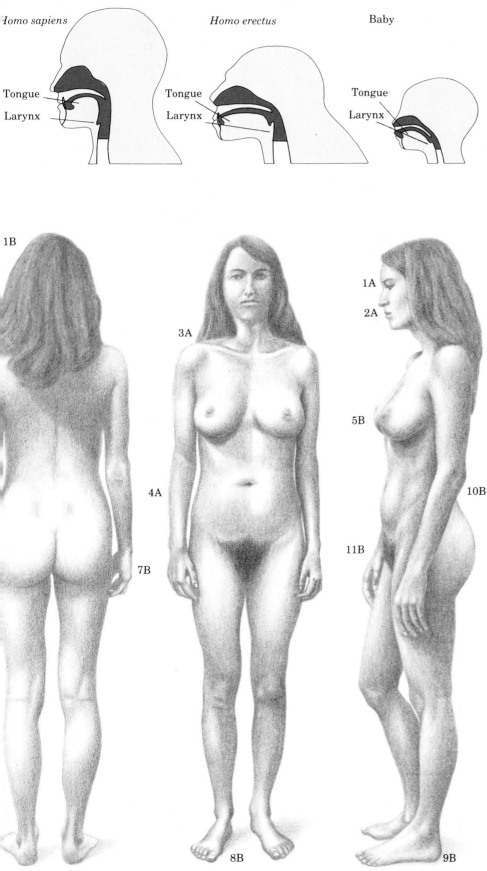

Homo sapiens   *Homo erectus*   Baby

Tongue   Tongue   Tongue
Larynx   Larynx   Larynx

1B
3A
1A
2A
5B
4A
10B
7B
11B
8B   9B

THE FOSSIL EVIDENCE OF MAN

**While it is possible to trace man's ancestry, although the fossil evidence
is incomplete, the exact process by which man has evolved is a controversial subject.**

When Charles Darwin wrote *The Origin of Species* in 1859, he avoided discussion of the origin of man, being content to note that, from his studies, 'light would be thrown on the origin of man and his history'. Twelve years later, in 1871 in *The Descent of Man*, Darwin relied on evidence of the comparative anatomy of man and apes – for example, the existence in man of rudimentary organs such as the appendix, and of embryological development – because he was obliged to refer to 'the absence of fossil remains seeming to connect man with his ape-like progenitors'.

The first authentic remains of fossil man were discovered in 1857 in the Neander Valley in Germany, but were attributed to those of a diseased modern man. The forehead was low and there were massive brow ridges, features previously considered to be characteristic of the apes. T. H. Huxley considered the bones of this early man the most ape-like known, but he stated that, 'In no sense can the Neanderthal bones be regarded as the remains of a human being intermediate between men and apes'.

Later the Dutch anatomist Eugène Dubois (1858–1940) became fired with the ambition to discover Darwin's 'missing link'. He gave up his university post and, unable to raise any financial support for an expedition to the Dutch East Indies, enlisted with the Dutch East India Army in 1887. In 1891, in Java, he discovered a skull cap and thigh-bone which he assigned to *Pithecanthropus erectus*, 'the erect apeman'. The name *Pithecanthropus* had been invented previously by the German zoologist Ernst Haeckel (1834–1919) for the hypothetical missing link between man and ape. These remains led to such controversy that Dubois eventually became a recluse because his interpretation of the fossils as belonging to an intermediate ape-man was not accepted. The general view was that they belonged to man, and indeed they are now designated *Homo erectus*.

In 1906 a complete Neanderthal skeleton was discovered at La Chapelle-aux-Saints in south-west France. The palaeontologist Marcellin Boule (1861-1942) described it; it was reconstructed with the head thrust forward and an ape-like shambling gait. This image has dominated the popular notion of the appearance of prehistoric man. It was not until 1958 that this conception of Neanderthal man was finally laid to rest. It was established beyond doubt that the original reconstruction was based on a skeleton severely distorted by osteo-arthritis. Thus there is no reason to believe that the posture of Neanderthal man was any different from that of present-day man.

In 1959 the anthropologist Louis

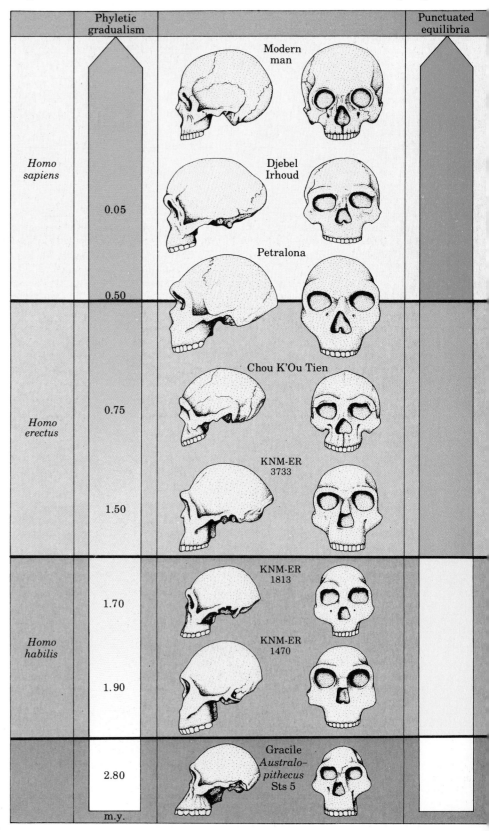

	Phyletic gradualism		Punctuated equilibria
Homo sapiens	0.05	Modern man / Djebel Irhoud / Petralona	
Homo erectus	0.50 / 0.75 / 1.50	Chou K'Ou Tien / KNM-ER 3733	
Homo habilis	1.70 / 1.90	KNM-ER 1813 / KNM-ER 1470	
	2.80	Gracile *Australo-pithecus* Sts 5	
	m.y.		

Leakey (1903–72) discovered 'Nutcracker man' (so called in honour of his large teeth), or *Zinjanthropus boisei*, now included in *Australopithecus*. The find was made at Olduvai Gorge, Tanzania. Since then, there has been a constant stream of major discoveries from Tanzania (Mary Leakey), from Lake Turkana in Kenya (Richard Leakey), and from Ethiopia (Johanson and Coppens). There have also been discoveries of great importance in Turkey, Hungary and Greece, in particular the Petralona skull from Greece, which sits exactly on the structural boundary between *Homo erectus* and *Homo sapiens*.

Even though there are now many documented remains of fossil man, the way in which they are interpreted varies and they have become the subject of heated disagreement among scientists. However, the facts cannot be disputed: the fossils exist and their sequence and relative ages are firmly established.

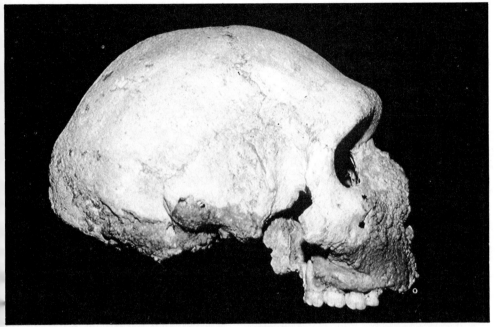

▲ The Petralona skull discovered in northern Greece in 1960. This skull is exactly on the morphological boundary between *Homo erectus* and *Homo sapiens* and is currently considered to represent the very first stage in the development of *Homo sapiens*.

▼ A reconstruction by the Russian fossil-restorer M. Gerasimov of the head of Rhodesian man. This advanced example of *Homo sapiens* lived 40,000 years ago.

◄ Sequence of fossil skulls showing the changes through time from the lightly-built gracile australopithecines to modern man. The diagram also illustrates two different interpretations according to the two current theories of evolution – phyletic gradualism (gradual change through time) and punctuated equilibria (sudden evolutionary changes followed by periods of stasis or equilibrium).

There is a sequence of types within *Australopithecus*, the gracile or lightly-built types being considered closest to the ancestors of man. The most advanced autralopithecines are usually classified as *Homo habilis*; the most advanced forms of these are *Homo erectus*; and the most highly developed of these became *Homo sapiens*. The traditional Darwinian view is that the evolution of man was a gradual process; if a complete fossil record could be found, it would reveal a series of finely graded human types, increasing in similarity to modern man as the geological sequence approached the present day. Morphological gaps between ancestral and descendent populations are attributed to the imperfectness of the fossil record. This interpretation has been described as 'phyletic gradualism', in marked contrast to the totally different concept known as 'punctuated equilibria' (see pp 82–3). This latter approach considers that the fossil record is fairly complete and that the gaps reflect real jumps in evolution. The promoters of this view have stated that there is no direct evidence for gradualism within any hominid category, and that each species disappeared looking much as it did at its origin.

A study in support of the gradualist view was published in the international magazine *Nature* in July 1981. It was by the anthropologist J. E. Cronin of Harvard University, USA, who, together with his colleagues, had surveyed all the known material of fossil man. Using modern dating techniques, these scientists were able to show that the sequence of types within a given species, such as *Homo erectus*, exhibited significant changes through time. Furthermore, as predicted by their model, there were several examples, such as the Petralona skull, which sat exactly on the boundary of two fossil species with intermediate characteristics between the two. Phyletic gradualism would suggest that, in the Middle Pleistocene, there would have been a species of man showing advanced *Homo erectus* features as well as primitive *Homo sapiens* features; the Petralona skull is an exact example of this. It does not matter whether it is classified as *Homo erectus* or *Homo sapiens*; its features place it on the boundary, and any firm categorization must be arbitrary. Refined dating techniques have been able to order the material on fossil man in the correct time sequence. The pattern that emerges is consistent with gradualism rather than punctuated equilibria, although the gaps in the fossil material remain.

ICE AGE MAMMALS

During the ice ages, mammals had to become more adaptable in order to survive.

The rise of mankind from the small, lightly built dryopithecines inhabiting the grasslands of Africa and Eurasia was, in geological terms, slow in coming. The earliest evidence of man comes from 3.6-million-year-old Pliocene rocks at Laetoli, Tanzania, in the form of footprints. But man's evolution has really occurred during the past two million years, in the Pleistocene Ice Age.

During the harsh and fluctuating conditions of the Ice Age, the mammal faunas underwent a fundamental change. Many groups evolved cold-climate forms, like the woolly mammoth and woolly rhinoceros. During the repeated advances of the ice-caps, vast regions of Europe, Siberia and North America were covered by tundra which, just as with grasslands, was capable of supporting vast herds of mammals. When the ice retreated and more temperate conditions prevailed, warm-dwelling mammals migrated

northwards. The history of the mammalian faunas of Eurasia in particular is that of warm- and cold-dwelling faunas alternately migrating north and south. Giant forms evolved: for example, horses, bison, aurochs or primitive oxen, deer, cheetah, wolves, mammoths and elephants were all larger during the Ice Age than their present-day counterparts; in Africa there was a giant warthog 3m long, twice the size of the modern one; an Asian rhinoceros was double the length of the surviving form; and a North American beaver was nearly 3m long, three times the size of the modern beaver. As the climate grew colder, many of the herbivores increased in size in order to conserve heat; a large body retains heat longer than a smaller one and is less susceptible to variations in temperature.

Although many of the highly specialized cold-dwelling fauna died out, most of the warm-dwelling mammals, such as deer, horses, cattle, pigs, hippopotamuses, cave lions, bears, hyaenas and elephants, flourished and simply retreated as the climatic zone moved. Some 100,000 years ago, during the last interglacial period, the area now occupied by Trafalgar Square in central London was inhabited by elephants, hippopotamuses, pigs, aurochs, bison, red and fallow deer,

carnivorous brown bears, hyaenas and lions.

The most outstanding feature of Pleistocene mammals is that their diversity is less than that of the early grassland radiations. Modern mammalian faunas are dominated by a

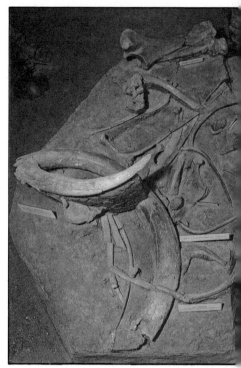

▲Bones of a woolly mammoth and a temperate climate straight-tusked elephant, which were found at Aveley, Essex, England.

▲The fauna of the Pleistocene Alaskan tundra, which was dominated by cold-loving mammals.

Extinct animals
1 Woolly mammoth 2 Ground sloth
3 Large-horned bison 4 Stag moose 5 Camel

6 Woodland musk ox 7 Woodland musk ox
8 Lion-like cat 9 Short- faced bear
Survivors, not now living in Alaska
10 Wapiti or elk 11 Musk ox 12 Yak 13 Horse
14 Saiga antelope 15 Badger 16 Ferret
Survivors still living in Alaska

17 Moose 18 Antlers of caribou 19 Dall sheep
20 Man 21 Beaver (lodge shown) 22 Alaska tundra hare 23 Brown lemming
24 Grizzly bear 25 Wolf 26 Lynx 27 Wolverin
28 Red fox 29 Arctic fox

few generalists. The most obvious one is man and it is often assumed that the impoverishment of the mammals is due to his direct intervention. Man may well have contributed to this state of affairs, but only marginally.

The Ice Age witnessed the

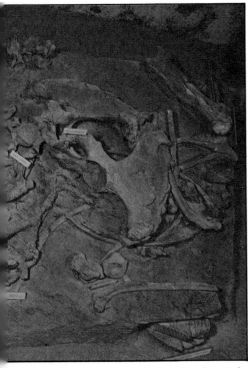

The skeletons date from an interglacial period in the Middle Pleistocene, when the climate was warmer than it is today.

establishment of a new evolutionary strategy: there were relatively few types of mammal, but those that flourished were highly adaptable. These were the sheep and cattle, cats and dogs, rats and mice. During the harsh fluctuating conditions that prevailed from season to season, there were no stable ecological niches for animals inhabiting temperate and higher latitudes; thus these animals had to be able to switch food preferences. Although they are familiar today because of their association with man, these animals had already established themselves successfully before man domesticated them. Man was another mammal that adopted this non-specialist food strategy.

The most successful of the herbivores are cattle, sheep and goats, none of which are specialized eaters, although sheep and goats are able to exploit more difficult terrain than cattle and crop vegetation closer. Pigs, although mainly woodland-dwellers, are sufficiently catholic in their food requirements to be highly successful too. Cats and dogs, which are happy to consume the flesh of any creature, continue to flourish and, being opportunists, are not averse to scavenging from man. Rats and mice have also had an evolutionary explosion.

There seems little doubt that the rise of man and the development of this new generalist strategy was a major consequence of the Ice Age. Although individual scientists had proposed the former existence of glaciers – and, indeed, had supported their contention with evidence – no appreciable advance was made until 1838. In this year Louis Agassiz proposed his theory, stating that a vast ice sheet had extended from the North Pole to cover most of Europe (see pp 32–3). The concept of a past ice age of such magnitude was sensational and the scientific community, including Buckland and Lyell, was almost unanimous in rejecting it. Although Buckland was a diluvialist, the Universal Deluge or simply icebergs, seemed unable to account for some of his observations. In 1840 Buckland invited Agassiz to visit the north of England and Scotland to study the drift deposits. Agassiz's findings converted Buckland to his ideas and managed similarly to convince Lyell. Lyell, Buckland and Agassiz all presented papers to the Geological Society of London showing evidence of the former existence of glaciers in England, Scotland and Ireland. This marked the beginning of the acceptance of the former existence of the Ice Age.

▲ North American Pleistocene mammals of temperate latitudes.
1 Megalonychid ground sloth 2 Mallard 3 Grebe 4 Extinct otter 5 Frog 6 Extinct catfish 7 Turtle 8 Extinct beaver 9 Sabre-toothed cat 10 Extinct weasel 11 Pelican 12 Gallinule 13 Goose 14 Extinct peccary 15 Zebra-like horse 16 Short-faced bear 17 Extinct mountain lion 18 Swan 19 Stork 20 Mastodon 21 Extinct llama 22 Cormorant 23 Extinct grison 24 Extinct rabbit 25 Extinct pronghorned antelope

THEORIES ABOUT THE ICE AGES

There have been many attempts to explain the ice ages, but the correct version was only accepted as recently as 1976.

The idea that there had been a period in the past when major regions of the Earth were covered by a vast ice sheet became accepted in the mid-nineteenth century. By the 1860s it was firmly established that there had been more than one ice age, and that the ice-caps had advanced and retreated across Europe several times. Four major ice ages were recognized as having been world-wide.

The former existence of ice ages posed the following questions. First, what caused the first ice age, and the repeated advances and retreats of the ice caps? Secondly, would the ice caps return to cover the whole of Europe? If they were to return, all the industrialized countries in the present temperate zone would probably be under millions of tons of ice. Alternatively, they would become almost uninhabitable cold deserts – apart from a brief summer season, notable mainly for the profusion of blood-sucking insects, as experienced today in the wastes of Alaska.

One of the first attempts to explain the origins of the ice ages was made by a French mathematician, Joseph Alphonse Adhémar. In 1842 he published a book, *Revolutions of the Sea*, in which he suggested that the ice ages might have been due to variations in the way the Earth orbits the Sun. In particular, he considered what is known as 'precession'. As the Earth revolves round the Sun, it rotates on its axis; but, rather like a top that wobbles, its axis of rotation prescribes a circle, termed 'axial precession'. This is caused by the gravitational pull exerted by the Sun and Moon on the Earth's equatorial bulge. Each season begins at a particular point in the Earth's orbit called a 'cardinal point'. Axial precession causes the four cardinal points to move slowly clockwise around the orbital path. At the same time, the Earth's elliptical orbit round the Sun moves more slowly in an anticlockwise direction. The two movements together cause the cardinal points to shift along the orbit; this is called 'precession of the equinoxes', which complete a cycle every 22,000 years. Today, winter in the northern hemisphere begins when the Earth is

close to the Sun; 11,000 years ago, winter began when the Earth was at the opposite end of the ellipse, furthest away from the Sun. Adhémar believed that, under the latter condition, an ice age occurred every 11,000 years, first in one hemisphere, then in the other. While this theory has proved to be erroneous, it was the first attempt to

relate the ice ages to astronomical events. James Croll (1821–90) a self-educated scientist, was the next person to put forward an astronomical theory. In 1864 Croll suggested that the real cause of the ice ages might be due to variations in the elongation of the orbit itself. This is known as the 'orbital eccentricity' and is due to the

▲ The Earth today (left) and during the last ice age (right). 20,000 years ago vast ice sheets covered about 28.5 million sq km of land that is free from ice today. In the Arctic and parts of the North Atlantic, the surface waters were frozen and the sea-level was approximately 105m lower than it is at present.

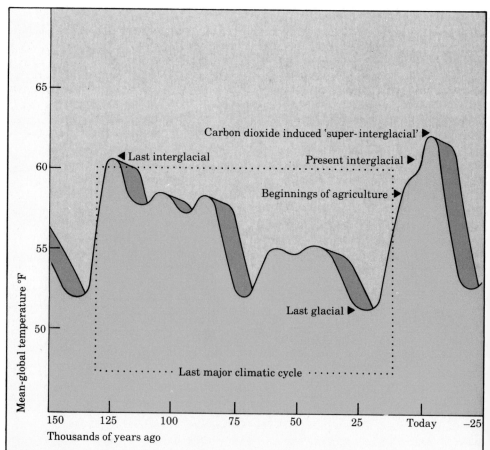

Climatic forecast for the next 25,000 years
▲ According to astronomical theory, the climate should return to the ice age 23,000 years from now. A 'super-interglacial' period may be caused by build-up of carbon dioxide in the atmosphere.

gravitational pull of the planets on the Earth's orbit. It had already been calculated that a cycle of eccentricity was completed every 100,000 years. In 1875 Croll published a book entitled *Climate and Time* in which he spelt out his astronomical theories in great detail. He was elected to membership of the Royal Society in 1876. However, his theory did not agree with the subsequent findings of geologists, as the last ice age ended 10,000 years ago, not 80,000 as Croll had predicted. By the turn of the century Croll's work was almost forgotten.

The next attempt at an astronomical explanation was made by the Yugoslav mathematician Milutin Milankovitch (1879–1958). In 1911, with a poet friend, he decided to develop a mathematical model that would allow the climate of Earth, Mars and Venus to be described for the present and the past, and for projection into the future. In this way he would reach an understanding of the ice ages.

Milankovitch's mathematical calculations lasted some 30 years. He based them on three variables: orbital eccentricity, axial tilt and precession. Variations in these factors had already been calculated by the German mathematician Ludwig Pilgrim. By 1930 Milankovitch had worked out the amount of solar radiation reaching Earth during each season in each of eight latitudes from 5° to 75°N. His next task was to find the mathematical relationship between the amount of solar radiation in summer and the amount of high land permanently covered in snow at the same latitude; he could thus discover how much increase in snow cover would result from changes in summer radiation. In 1941, at the age of 63, Milankovitch completed his mathematical theory of radiation. He had worked out how the ice sheets responded to the changes in solar radiation at different latitudes, and his graphs showed the latitudes of ice sheets over the past 650,000 years. Although at first welcomed enthusiastically by geologists, Milankovitch's theory seemed to be contradicted by results gained from the carbon-14 radiometric dating technique. By 1965, his astronomical theory had few supporters.

The matter was finally resolved in 1976 by the study of deep-sea cores from the floor of the Indian Ocean, which proved Milankovitch correct. Microscopic marine fossils deposit a skeleton of calcium carbonate: at higher temperatures the lighter oxygen isotope (Oxygen 16) is included in the calcium carbonate, while at lower temperatures the heavier Oxygen 18 is incorporated. The ratios of these isotopes can be used to calculate the amount of ice which existed at various stages in the Earth's history. During the past 500,000 years there is evidence of a repeated cycle in which the ice age recurred: 100,000, 43,000, 24,000 and 19,000 years ago, which matched the astronomical cycles of orbital eccentricity (100,000), axial tilt (41,000) and precession (24,000 and 19,000). In 1976 J. D. Hays, J. Imbrie and N. J. Shackleton wrote an article in the magazine *Science* called, 'Variations in the Earth's orbit; pacemaker of the ice ages', which has led to a prediction of a return of the next ice age.

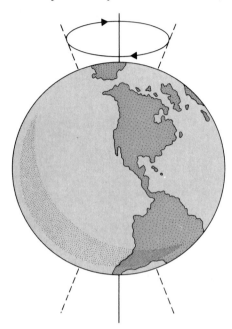
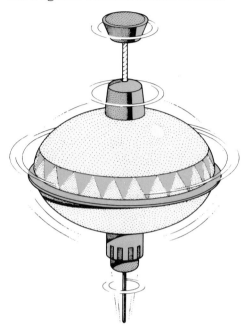

Axial precession
▲ The Earth rotates on its axis as it revolves round the Sun. The axis of rotation moves slowly in a circle, rather like a top that wobbles; it completes one revolution every 26,000 years.

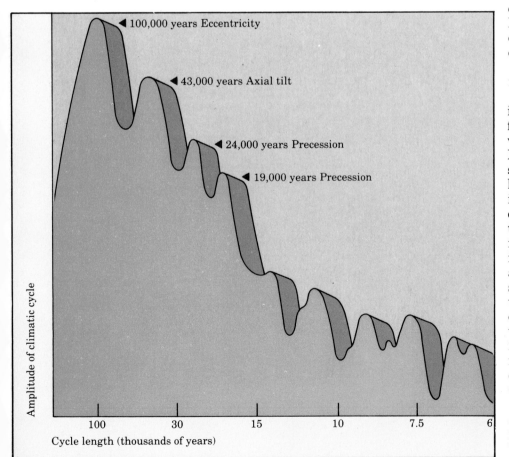

100,000 years Eccentricity

43,000 years Axial tilt

24,000 years Precession

19,000 years Precession

Amplitude of climatic cycle

Cycle length (thousands of years)

▲ The graph shows the relative importance of different climatic cycles in the isotope record of two Indian Ocean cores. This confirmed many predictions in the astronomical theory of the ice ages.

HOW EARLY MAN HUNTED

Early man's hunting techniques relied on his control of fire and his invention of the all-purpose tool, the Acheulian hand axe.

While early man's advance was related to his ability to respond to the changing and generally adverse conditions of the Ice Age, this sudden expansion does not appear to coincide with any particular event among the climatic cycles that marked the Ice Age. Instead it seems to have been related to man's hunting technique and weaponry. The key factor was the development of the Acheulian bifaced hand axe about a million years ago, which was the basic all-purpose tool of *Homo erectus*, and ideally suited for butchering large animals.

In conjunction with advanced tools, man also acquired mastery over a natural phenomenon which was greatly feared by all other animals: fire. The ability to make fire gave man a tool of immense power that set him apart from the rest of the living world. Until recently the first evidence of man-made fire came from a hearth some 70,000 years old, in a cave at L'Escale in southern France. However, at the end of 1981, evidence was found of a man-made fire 1,400,000 years old, associated with the simplest stone tools and australopithecine remains from near Lake Baringo, Kenya. Fire provided a sure protection against enemies or predators; it also provided warmth and a means by which meat

could be cooked. This not only made the meat more palatable, but also made more of the nutrients directly accessible. Indeed, at Chou K'ou Tien near Peking, China, which is associated with the remains and tools of *Homo erectus*, there are ash layers 6m deep and the charred bones of deer, sheep, horses, pigs and cattle.

Fire in its domestic setting was crucial to man's advance, but it also provided one of the most effective methods of hunting. Evidence of this was found at Ambrona, some 150 km north-east of Madrid in Spain. Here, in sediments about 400,000 years old, remains of straight-tusked elephants had been collected at the beginning of the century, but no systematic excavations had been undertaken. In 1960 F. Clark Howell of Chicago University revisited the old excavations. In three field seasons he uncovered the remains of over 50 elephants and some other animals. There were no human remains or signs of habitation, but there was substantial evidence of hunting and butchering. It is from this site that much of our knowledge of early man's hunting techniques has been reconstructed.

From fossil plant pollen it is evident that 400,000 years ago central Spain was exceedingly cold, experiencing tundra conditions in a region of permafrost. The alternate

freezing and thawing that took place annually pushed up the land in places to produce hills with areas of muddy ground in the hollows. The evidence of fine layers of carbon and charcoal scattered across the valley indicates that the low vegetation cover had been repeatedly burnt. It is reasonable to infer that this was done by man to drive herds of elephants into the quagmires in the valley bottoms. There the trapped elephants would have been

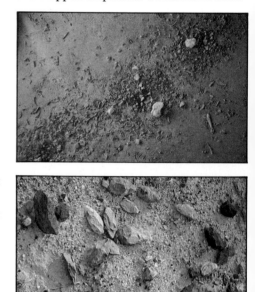

killed by smashing their heads with stones and then spearing them. Pieces of fossil wood were preserved in the sediments in the shape of sharp fire-hardened points, establishing that *Homo erectus* wielded spears.

Due to the meticulous excavation of F. Clark Howell, these elephant bones have revealed an enormous amount of valuable information. It emerged that the elephants were driven into the soft ground and there

◄ Three hearths at Pincevent, near Paris, France. The hearths are 12,000 years old and were in one Palaeolithic hut.

▲ A hut at Terra Amata, southern France, in the process of reconstruction. It is based on the excavations of the 400,000 year-old-site.

◄ *Homo erectus* hand axes from Olorgesailie, Kenya. The bones of over sixty giant baboons were found with the axes giving clear evidence that they were used for hunting.

▼ A reconstruction of the hunting techniques used at Ambrona, Spain, 400,000 years ago. Elephants were driven by fire into the swamps. An elephant skeleton with just one side preserved has been found – this shows that the animal was partly buried and the hunters were able to remove only one side.

killed. Once the animals had been killed, the carcases were cut up. The flesh and some of the bones were then taken to firmer ground, where they were cut up into smaller pieces and cooked. There are numerous hearths with clusters of burnt and broken bones. It is also clear that at these sites people were making tools for the task of cutting up the elephant meat. Such a massive hunt would have provided food for a large number of people and it is likely that several separate hunting groups co-operated. When the piles of remains were compared with each other, it became clear that each one contained examples of all the different animals that had been killed and butchered. Such a distribution could not have occurred by chance. The hunters must have divided their spoils equally among the different groups; this is characteristic of primitive hunter-gatherer peoples today.

A further insight into the life of *Homo erectus* came in 1965 when the French palaeontologist Henri de Lumley noticed some stone tools which had been uncovered on a building site in Nice. The richness of the material that began to be found at the site, now known as Terra Amata, resulted in de Lumley's being given from January to July 1966 to rescue what information he could before the building continued. The excavations revealed a series of 21

separate levels of habitation, which had been occupied only during late spring and early summer, about 400,000 years ago. Only simple shelters had been built. These primitive houses were made of young trees and saplings fixed into the ground. There was a hearth inside and part of the interior was used for cutting up meat. The people that occupied this camp site slept on hides, which suggests that they wore clothes. The most frequently hunted animals were red deer, elephant, boar and ibex, but oysters, limpets and mussels, as well as birds, turtles and fish, were also consumed. At the corner of the shelter were human faeces and fossilized pollen, mainly from flowering broom, which scatters its pollen in late spring; as no pollen from any other time of the year was present, the hunters could only have been on the site for a short period each year. This French Riviera camp seems to have been the spring location of a group of hunters that must have had a regular pattern of annual migration.

The only known site which was inhabited by *Homo erectus* as a permanent base was at Chou K'ou Tien; these habitations were in caves on the hills, overlooking the flat plains where game abounded. Red deer made up some 70% of the food items. Among *Homo erectus* in Chou K'ou Tien there is a hint that they indulged in cannibalism, as the skulls seem to have had the foramen magnum, at the base of the skull, deliberately widened in order to extract the brains. It is unlikely that people were eaten for meat, as food was in abundance; it is more likely that cannibalism was part of a ritual (see pp 172–3).

Graham Allen.

TOOLS

The increasing sophistication of prehistoric tools is evidence of early man's development.

Man was initially defined as 'the tool-maker', but this definition has had to be altered because other vertebrates modify natural objects for use as tools and hence would fall into the same category. Man is thus considered unique in the living world by virtue of his capacity for abstract thought. Since it is difficult to find evidence of this in the fossil record, man's tools are distinguished as being made for future use, as opposed to the *ad hoc* tools of other animals.

Flint hand axes of the type now associated with *Homo erectus* were first discovered over 3m below the surface in a brick pit at Hoxne in Suffolk, England, by John Frere (an ancestor of Mary Leakey) in 1797. He claimed that 'they had been used by a people who had not the use of metals' and went on to suggest that they were of 'a very ancient period indeed, even beyond that of the present world'. No one took much notice of these astute observations. Although flint implements were found together with bones in the early nineteenth century, it was not believed that the tools were the work of 'antediluvial' man.

In 1846 a fierce controversy broke out when Boucher de Perthes, a French customs officer and amateur archaeologist, announced that he had discovered flints worked by man in ancient gravels. He wrote, 'In spite of their imperfection, these rude stones proved the existence of man as surely as a whole Louvre would have done.' Boucher de Perthes' views found little support from the scientists of his day. But in 1859 two British scientists, Professor Joseph Prestwich and John Evans, visited France and examined the finds; they returned convinced that Boucher de Perthes had indeed uncovered genuine evidence of prehistoric man.

At the end of the last century and the beginning of the present, much debate took place over primitive flint tools found in Kent and later in East Anglia, England. These were termed 'eoliths' and were believed to represent the very beginnings of tool-making by man. However, the eoliths are in fact flints chipped without the intervention of man. Most seem to have been produced by friction caused by soil moving down-slope under periglacial conditions.

Controversy still continues over the tools of australopithecines. Professor Raymond Dart interpreted numerous broken bones associated with the remains of *Australopithecus* as having been modified for use as weapons and tools. It has always seemed reasonable that mankind's precursors would have used the remains of their food as the most readily available instruments, and there has been a continuous tradition of bone culture throughout man's history.

However, other scientists have been able to show that the fracturing of the bones that Dart examined can be replicated in African villages, near water holes, where goats trampling bones cause them to break in a comparable way. Some bones, when

Primitive

▲ Acheulian flint hand axes used by *Homo erectus*.

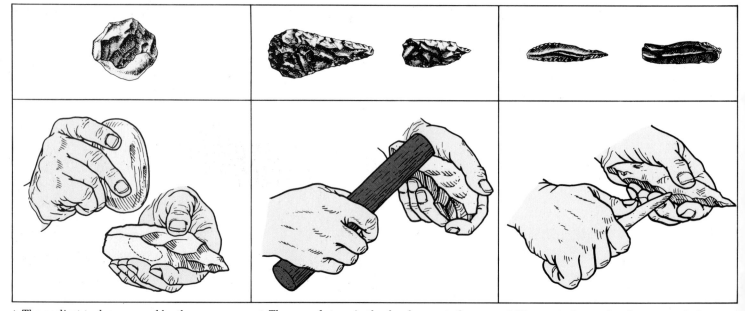

▲ The earliest tools were used by the australopithecines. These hand axes were simply pebbles which were chipped on one side by a larger stone to produce a sharp cutting edge.

▲ The second stage in the development of tools was the general-purpose tools invented by *Homo erectus*. They were made from a flint core which was chipped.

▲ The most advanced tools were made from flint flakes which were then pressure flaked to produce fine tools. The use of pressure flaking indicates the precision grip used by *Homo sapiens*.

eroded by stomach juices in hyaenas, also resemble examples of Dart's osteodontokeratic (bone-tooth-horn) culture. The skull fractures of baboons or australopithecines, which Dart claimed had been made by a right-handed being wielding the long bones of antelopes, could have been caused by rock falls.

Dart's case may have been overstated and some of his interpretations are impossible either to substantiate or to refute; for example, there is no way of telling whether or not a jawbone was used to tear open carcasses. However, at the cave living-site of Makapansgat, South Africa, bones were found which could not have been fractured by animals trampling round a water hole because of the situation of the cave; and the number of double fractured skulls discovered indicates that they could not have been caused by chance rock falls alone. Makapansgat seems to confirm Dart's interpretations.

Whatever the views on Dart's osteodontokeratic culture, there seems to be general agreement as to the authenticity of the pebble tools associated with the australopithecines in East Africa. The Oldowan chopper tool is a hand-sized rounded pebble which was struck by a larger stone so that it became chipped along one edge; the smooth rounded part could be easily held in the hand. The authenticity of these artifacts is proved by the fact that they are always found at sites of man-made shelters and structures. Strangely, many of them look much less impressive than the discredited eoliths.

The first general-purpose tool invented by *Home erectus* was the hand-sized axe, in which the flint was shaped by chipping off flakes from the outside. This was done with a hammer stone and produced a tool made from the flint core. The edge of the crudely fashioned hand axe could be further flaked to produce a cutting edge and such tools were extremely effective for butchering large mammals. The butchering sites described by Mary Leakey in Tanzania have revealed biface hand axes, some of which served as scrapers; there were also round stone balls which Leakey considers would have been used like the South American bolas, whirled round the head on thongs of sinew and hurled at the legs of animals to bring them down.

With later *Homo erectus*, the biface hand axes were less crudely shaped and served as meat-cleavers, with side-scrapers for cutting and skinning carcases. More advanced stone tools were made in a different way: instead of fashioning a single tool from a piece of flint, a series of flint blades were flaked off the core. This type of blade tool was produced by hitting the core using a bone or wood punch, which in turn was hit with the hammer stone. Longitudinal blades flaked off, giving a characteristic grooving to the core stone. The razor-sharp flakes were then converted into a variety of fine tools by pressure-flaking. This highly skilled task used a piece of bone or hard wood to chip off minute flakes along the edges of the flint, first by working along one edge, then by turning it over and working along the opposite side of the same edge. The development of fine pressure-flaking indicates the evolution of the precision grip and is a sign of modern man, *Homo sapiens*.

Advanced

▲ A primitive pebble tool from Olduvai Gorge, Kenya, over a million years old.

◀ Simple stone tools from Karari, Kenya, which were used by the australopithecines. These are 1.5 million years old and include lava and quartz choppers.

RITUAL AND ART

Neanderthal man had burial ceremonies and various cults. His art may have been related to magic, but also shows a sense of humour

The development of the flake-tool industry is associated with Neanderthal man, now classified as belonging to our own species *Homo sapiens*. More is known of the way of life of the Neanderthals than of *Homo erectus*, especially from several sites in France where the Neanderthals built shelters of saplings covered in hides within the mouths of caves. They cooked and even invented the hot-plate by building a fire on flat stones, then putting out the fire and placing meat on the hot stones. They also sewed their clothing of animal skins.

One characteristic of the Neanderthals was their interest in ochre-red pigment. This pigment must also have had an important meaning for *Homo erectus*, as some 75 pieces of ochre were found at Terra Amata. Since *Homo erectus* was a hunter-gatherer, the ochre must have been significant enough for him to carry it on his expeditions. Hand-sized lumps of pigment are also associated with traces of colour inside bones and in hollow rocks at Neanderthal sites. The colouring was probably for marking the body for ritual purposes. In view of the body decoration of living primitive peoples, this does not seem an unreasonable supposition, although there is no direct evidence for it.

The Neanderthals had various cults; some of the most striking evidence of them is found at Neanderthal burial sites in Europe and Asia. The ceremonial for the dead included a fire lit on top of the body. In one grave a hunter has been found buried with a bison leg on his chest and surrounded by broken bones and flint tools. In 1960 a Neanderthal grave at the back of a cave was excavated in Shanidar, in the Zagros Mountains of northern Iraq. A routine pollen analysis revealed vast amounts of pollen in patches, showing that bunches of flowers must have been placed in the grave; they included bunches of grape hyacinths, hollyhocks, groundsel and cornflowers. The nature of the burials suggests that some kind of belief in an after-life had arisen by this time. The care with which the dead were buried and the sign of respect shown by placing flowers on the body seem similar to modern customs.

The average lifespan of a Neanderthal seems to have been 30 years for a man and 23 years for a woman. A number of Neanderthal remains include exceptionally old individuals, such as the La Chapelle-aux-Saints skeleton of a man who lived until the age of about 60; he was toothless, apart from the lower incisor teeth, and had suffered from severe osteo-arthritis. Other people had been severely handicapped, such as a 40-year-old man who had had the use of only one arm from birth. The fact that disabled individuals had a place in the community and lived to such ages suggests a highly developed and caring society, something for which there is no evidence as yet among earlier men.

The graves show that Neanderthals were killed by spears, and wounds frequently on the left side imply fights between right-handed people. Although Professor Dart has suggested that australopithecines killed

▲ An early 'Venus' holding a horn carved in limestone from Laussel, south-west France.

▲ Pelvis of woman from the HaYonim cave, Israel, which still carries a girdle of fox teeth.

▲ Two bison, modelled in clay 15,000 years ago, on the floor of Le Tuc d'Audoubert cave in southern France.

▲ A carved stone bison from southern France.

one another, the evidence that Neanderthals did so is unequivocal.

Evidence of cannibalism has been found. Considering the abundance of animals and the small amount of flesh available from an average human body, (less than 5 kg), cannibalism was probably a religious ritual. In 1899, at Krapina in Yugoslavia, the mutilated remains of 20 people, men, women and children, were discovered with the skulls smashed, the long bones split and also charred. In Solo, Java, eleven skullcaps were found together with the faces and jaws smashed, but no sign of any other remains. Most of these skulls show that the foramen magnum had been opened by stone or wooden implements. Present-day cannibals use the same methods in order to extract the brains from skulls. There are many examples that point to cannibalism among the Neanderthals and there is some evidence of the same traits among *Homo erectus*. Ritual

cannibalism is fairly conclusive evidence of some kind of religion having become established.

The Neanderthals are well known for their cave-bear cult, presumably connected with the fact that they competed with the bears for living sites. Bear skulls have been found in natural niches in cave walls, occasionally with a limb-bone thrust through the arch of the cheek-bones. At Montespan in the French Pyrenees, a life-size clay sculpture of a headless bear from the Upper Palaeolithic has been discovered in a cave.

The first advanced members of our species are commonly known as Cro-Magnon man. It used to be thought that Neanderthal man was a separate side branch that was wiped out by invading modern or Cro-Magnon man. The current view includes both Neanderthal and Cro-Magnon under *Homo sapiens*. Cro-Magnon man seems to have spread throughout most of the world, reaching Australia and South America. In Eurasia he evolved many new ways of living, the best documented being that of the mammoth-hunters of the European plains in what is now the Ukraine. The mammoth-hunters built their homes of mammoth bones covered by hides and dressed in clothing made of hare skins. In 1981 it was discovered that they had constructed a set of musical intruments from mammoth bones.

Perhaps the most memorable aspects of their culture were the carved bone and ivory representations of the animals and people around them. They painted examples of the living fauna so that we know exactly what the woolly mammoth and rhinoceros looked like. Many of their representations are considered to be some kind of sympathetic magic: by portraying an animal, its spirit is captured and the hunter has a certain power over it. Many of the representations of human beings are evidently connected with fertility, like the voluptuous ivory or stone sculptured 'Venuses' found around the Mediterranean. Examples of ornaments, such as necklaces and bracelets, have also been found. There are nearly life-size portrayals of youthful female forms – and finally there are what appear to be jokes. Even 20,000 to 30,000 years ago there is ample evidence that man had a sense of humour. This can be seen, for example, in the pictures of a bison being transformed into a woman. Palaeolithic or Stone-Age art has as many facets as more sophisticated art, and does not always serve a specific religious function.

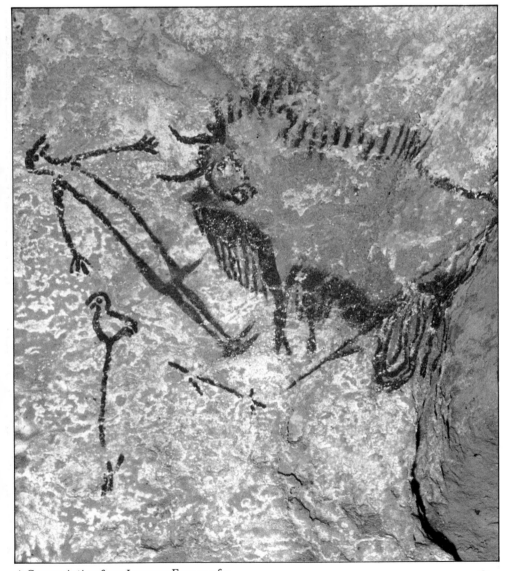

▲ Cave painting from Lascaux, France, of a bison, the figure of a man and a bird symbol.

▶ A drawing based on a series of cave paintings showing a bison gradually changing into a woman – evidence of Palaeolithic humour.

THE NEOLITHIC REVOLUTION

The Neolithic revolution occurred when man gave up his hunter-gatherer existence and settled in one place to cultivate crops and rear animals.

A major advance in man's history was described by the British Professor V. Gordon Childe as the Neolithic revolution; it involved the cultivation of plants and the domestication of animals some 10,000 years ago. Between the hunter-gathering economy and the domestication of animals, there was an intermediate stage when wild animals, particularly ungulates, were herded together in much the same way as the Lapps today herd reindeer. It was but a short step from herding to animal husbandry, where the animals were protected and confined within a circumscribed area and could breed in safety. Whereas in the wild it was mainly the very young and old animals that were killed in the hunt, domesticated animals could be slaughtered as soon as they reached peak condition.

The Neolithic revolution began in the Middle East along the flanks of the mountains bordering the 'Fertile Crescent', the arc from the Tigris-Euphrates valley to that of the Nile. Excavations at Jarmo in north-eastern Iraq have revealed a hunter-gatherer way of life combined with that of a settled agricultural community. The diet seemed to comprise mainly gazelle and wild ass, but also included snails and various nuts and fruits, in particular acorns. There were about 24 mud houses, which must have been inhabited by some 150 people. It was clearly a settled community, which lived here throughout the year.

It is in just this region, along the flanks of the Fertile Crescent, that wild wheat and barley still flourish. Evidence of sickles with small flint blades and pestles and mortars has been found at Jarmo. A small family armed with a Neolithic sickle could have collected sufficient grain during the three-week harvest period to last a full year. Grain was easy to store and could be ground down to make porridge; if allowed to ferment, it produced beer. It is not difficult to envisage the transition from collecting grain to sowing it deliberately in chosen places. Straw was a useful by-product of the grain harvest; when

mixed with mud, it made bricks for building. The availability of grain enabled human beings to establish permanently settled communities as there was no need to search for food.

At much the same time in Iraq the first animals were domesticated, in particular the wild sheep. Evidence of domestication comes from the remains of the animals, which are all of young

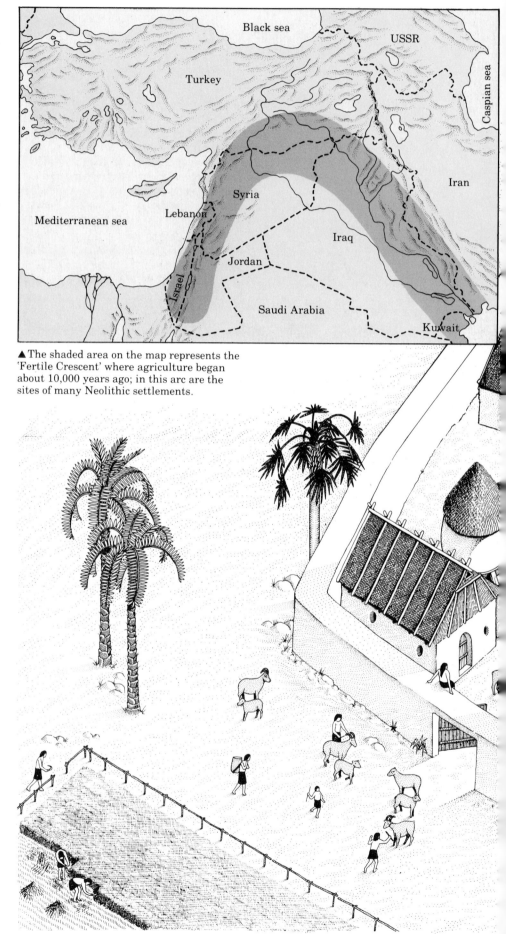

▲ The shaded area on the map represents the 'Fertile Crescent' where agriculture began about 10,000 years ago; in this arc are the sites of many Neolithic settlements.

mature animals of the same size; they had obviously been kept until they reached their peak condition and then butchered. Curiously, during domestication the cross-section of the horn cores of sheep and goats changed. In the domesticated forms one side became flattened, whereas the horn was convex on both surfaces in the wild types. There is no explanation for this difference. Sheep in Iraq were domesticated about 10,500 years ago, followed by goats in Iran 9,500 years ago, pigs in Turkey 9,000 years ago, and cattle in Greece 8,500 years ago.

One of the best-documented Neolithic localities is Jericho in Jordan, the site of a small fishing hunter-gatherer settled community near a spring. This site has always been considered the birthplace of Neolithic culture, as the first evidence, in the form of buildings, pots and cloth, was found here. Jarmo, however, seems to represent an intermediate stage of development.

The key feature of the settled communities of the Neolithic, with their cultivation of crops and husbanding of animals, was that a small number of people could provide food for the whole community. This left people free for other specialized tasks. Entirely new occupations and crafts came into existence, like building houses and food stores, weaving cloth and making pots. Thus the Neolithic marks the beginning of civilization, a word derived from the Latin, *civitatem*, meaning town.

In the early Neolithic, man's main concerns were the production of crops and the breeding of animals. Understandably, his religions were dominated by Mother Earth and the cult of the mother goddess. In many present-day primitive societies such concepts are still dominant, especially as the main cultivation is usually done by the women. Settled farming communities in small villages marked a fundamental change in man's way of life. Whereas the hunting community celebrated the virtues of the hunter, a settled community required a wider variety of skills. Since food was abundant, little time needed to be spent collecting it.

Such a state of affluence leads to rapid population growth and eventually to serious pressure on resources, but in the first instance it leads to an increase in the individual's expectations. People start wanting luxuries, that is, objects from outside their immediate vicinity. Trading begins and before long the entire train of modern human history is set in motion. Once there are extra resources, trade and trade routes, men can acquire wealth by raiding; they can achieve power and status by guarding the trade routes. Armies and organized religion arise to maintain the *status quo*. Villages develop into towns, as these provide greater security. Settled communities, by thriving on a diversity of skills, provide a basis for growth, development and expansion.

▼ The way of life in a settled Neolithic community. The early farmers reared animals for meat, milk and skins; wheat and barley were cultivated; houses and grain stores were built , pottery and textiles were made.

THE PILTDOWN HOAX

Piltdown man – 'the first Englishman' – turned out to be a fraud, but the story behind it has never been fully revealed.

More has been written about Piltdown man than about any other fossil man. It was found by amateur palaeontologist Charles Dawson and named *Eoanthropus dawsoni* in his honour. Piltdown man was discovered in a gravel pit at Piltdown, Sussex, England, and was hailed as the most important find ever made. The fragments of skull showed that the braincase was similar to that of modern man, but the jaw was ape-like, although critical parts of the lower jaw – namely, the anterior part where the two halves joined, the canine teeth and articulation – were missing. At the meeting at which the discovery was announced, the eminent zoologist Sir Ray Lankester stated that 'he did not consider it certain that the lower jaw and the skull belonged to the same individual'. Professor Waterston also found it difficult to accept that all the bones came from the same individual; he noted that the skull was 'human in practically all its essential characters', while the lower jaw 'with equal clearness resembled, in all its details, the mandible of a chimpanzee'.

Until this discovery, all important fossil man finds had been made in France and Germany, much to the distress of the British. Indeed, it was known that the head of the Geology Department of the British Museum (Natural History), Dr Arthur Smith-Woodward, had ambitions to discover the remains of the first man in England. The saga began in early 1912, when Dawson showed Smith-Woodward the fragments of a human skull he had discovered at Piltdown. At week-ends later that year the two men sorted through gravels, helped by a young Jesuit priest, Father Pierre Teilhard de Chardin. Smith-Woodward felt that the skull fragments were of little value without the jaw; on one of their week-end excavations Dawson's hammer flicked up the sought-for mandible. It was an ape-like jaw, but the critical canine tooth was missing.

During excavations the following year, Teilhard de Chardin sat resting on a pile of gravel which had just been sieved by Dawson and Smith-Woodward. After about five minutes' casual probing with his fingers, he picked out the hoped-for canine. As Smith-Woodward remarked, 'We were incredulous.' During the first season, the men found a number of primitive

flint tools and various fragments of fossil bone and teeth, including mastodon, elephant, beaver, red deer and horse. In 1914 they discovered a unique bone implement in the soil under the hedge where they normally had their lunch. As Smith-Woodward noted, 'It had been shaped by man and looked rather like the end of a cricket bat.' Smith-Woodward later wrote a full account of the Piltdown story, *The Earliest Englishman*, completing it the day before he died in 1944.

Piltdown man engendered tremendous enthusiasm. It also confirmed the anthropological establishment's belief that man's brain had evolved first, then the ape-like jaw

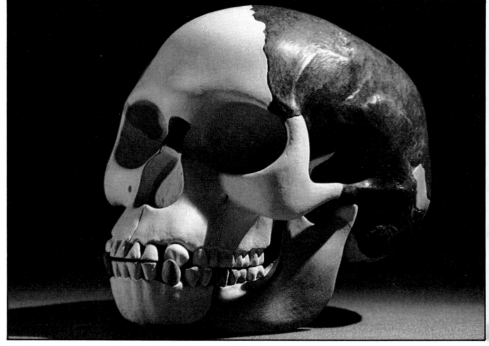

▲A reconstruction of the Piltdown skull. The brown parts are the actual material which was found at Piltdown; the remainder has been reconstructed in plaster.

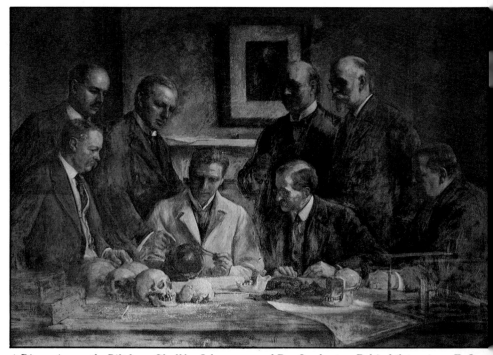

▲*Discussion on the Piltdown Skull* by John Cooke R. A. Sitting at the table (l. to r.) are W. P. Pycraft, Arthur Keith, A. S. Underwood and Ray Lankester. Behind (l. to r.) are F. O. Barlow, Grafton Elliot Smith, Charles Dawson, and Arthur Smith–Woodward.

and lastly man had walked upright.

In 1925, in *Nature* magazine, Professor Dart announced his discovery of the Taung skull, *Australopithecus*. He described it as having a brain the size of a chimpanzee, a man-like jaw and an upright gait and posture, the very antithesis of Piltdown. The Piltdown experts, the anatomists Sir Arthur Keith and Professor G. Elliot Smith, and Smith-Woodward, all responded immediately in *Nature* by roundly attacking Dart's interpretation. Only one major figure came to Dart's assistance – Professor W. J. Sollas of Oxford University, a bitter enemy of Smith-Woodward. All subsequent discoveries confirmed Dart's thesis and Piltdown man began to appear more and more anomalous.

Dr J. S. Weiner of Oxford University was the first scientist to consider seriously the possibility that Piltdown man's lower jaw and canine tooth might be modern ape faked to resemble fossils. Subsequently a detailed re-examination proved that the lower jaw came from a 500-year-old orang-utang; the jaw had been stained and the teeth artificially filed down. Further study revealed that the flint tools were fakes and that the fossils of the entire fauna of mammals had been deliberately placed there. The greatest hoax ever perpetrated in the history of palaeontology had finally been exposed after 40 years. Unfortunately no serious attempt was made to discover either the motive or the culprit. In 1955 Dr Weiner wrote a book, *The Piltdown Hoax*, which directed suspicion at Charles Dawson.

Matters took a surprising new turn in 1978, with the discovery of a tape-recording by Professor J. A. Douglas of Oxford University. Just before his death, Douglas had accused his predecessor, Professor Sollas, of being the probable instigator of the entire hoax; he supported his contention with a list of suspicious incidents. It appeared that there may have been a conspiracy to deceive Smith-Woodward. It is true that Smith-Woodward had been a very unpopular man. He was tremendously arrogant, lacked imagination and was not known for a sense of humour. One of his main enemies was Dr Martin Hinton of the British Museum. Hinton had an argument with Smith-Woodward in 1912 and was subsequently banished from the Geology Department. Circumstantial evidence, such as Hinton's access to the Everett Collection of medieval orang-utan material, indicates that he may have been directly involved in the hoax, as possibly was de Chardin. A number of young men on the staff of the Natural History Museum are known to have disliked Smith-Woodward. It is likely that they concocted the hoax with Sollas's encouragement. Sollas had been privy to Smith-Woodward's falling for another hoax which Sollas subsequently exposed; this was the Sherborne horse's head, an engraving done by some schoolboys which Smith-Woodward had described as a piece of prehistoric carving.

Unfortunately, the Piltdown hoax came at a moment when it engendered such a flood of patriotic fervour that scientists critical faculties went into abeyance. The theatrical 'discoveries', made in exactly the order that Smith-Woodward wanted, should have aroused suspicion and the crudely carved cricket bat should have demonstrated that the whole affair was a joke. It was not until Douglas's tape was discovered that Sollas was implicated. Hinton's role was revealed by Dr L. Harrison Matthews, an eminent zoologist, who studied Kenneth Oakley's files on the Piltdown man. (Oakley had spent almost a lifetime working on the 'fossil'.) In 1981 Matthews published much of the story in the magazine *New Scientist*.

▲A bone implement from Piltdown which looked like a cricket bat, an appropriate tool for the 'first Englishman'.

▲The *Illustrated London News* published this reconstruction of *Eoanthropus dawsoni* immediately after the Piltdown discoveries. The drawing was based on scientists' advice and shows some of the fictitious Piltdown 'fauna' in the background.

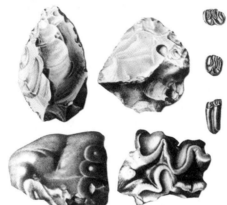

▲Fossil remains and artefacts which were also found at Piltdown and described by Smith-Woodward. The tools had been doctored and none of the fossils genuinely came from the site.

▲A picture published in the *Illustrated London News* of 4 January 1913 showing Charles Dawson (left) and Arthur Smith-Woodward searching for remains of Piltdown man at the site.

CHAPTER 6

HUNTING FOR SIGNS OF
THE EARTH'S EVOLUTION

During the 1960s geology witnessed a revolution
in the understanding of the dynamics of the Earth's
internal processes. With the concept of 'plate tectonics'
suddenly hitherto unexplained observations made sense.
The same fossils are found in different continents
because the continents were once joined. The rocks of
the ocean floors are relatively young because the ocean
floors are continually being generated and recycled.
Mountains were formed in mountain-building episodes
related to the collision of moving plates of
the Earth's crust. The unifying theory of plate tectonics
has accomplished for the Earth sciences what Darwin's
concept of evolution did for the biological sciences.

THE IDEA OF CONTINENTAL DRIFT

It was difficult to accept that continents had drifted, because there was no evidence of a mechanism which could move land masses.

The idea that the continents had not always been in their present positions arose from a variety of observations. The first was the apparent fit of the coastlines of South America and Africa. As long ago as 1620 Sir Francis Bacon had commented upon this. Other commentators discussed the possibility that the two continents were once united and attributed their subsequent separation to the Biblical Flood. In 1858 Antonio Snider-Pellegrini was the first person to use geological evidence to support the theory that the Old and New Worlds had split apart due to a catastrophic event caused by the Flood. However, this was the last example of the Flood being cited because, by the mid-nineteenth century, such notions of catastrophes were discounted by all serious scientists.

It was the work of the German astronomer and meteorologist Alfred Wegener (1880–1930) that caused the theory of continental drift to be taken seriously. In 1910 Wegener was impressed by the apparent fit of the two margins of the Atlantic Ocean, but it was not until the following year that he really began to consider the available geological evidence. He came across palaeontological evidence of a connection between Brazil and

southern Africa in the form of the small freshwater fish-eating reptile *Mesosaurus*, found in similar Permian deposits on both sides of the Atlantic, but nowhere else in the world. He published his theories on the movement of continents in 1912. In 1912–13 he was working in Greenland; then with the outbreak of the First World War he was called up into the army. He was twice wounded and during his convalescence he expanded his earlier papers into a book. The third edition was translated into

English and published in 1924 as *The Origin of Continents and Oceans*.

This started a bitter controversy. Wegener, who was not a geologist, was challenging the long-established views on the origins of the continents. The Earth was believed to have begun as a molten ball and, as it solidified, the lighter continental materials which were more granitic had floated to the top, while a denser basaltic layer had formed underneath – the ocean bed. This was known as 'sima' because it is composed of silica and magnesium and

▲ Rocks at Irai, south India showing striations caused by glacial action. Evidence like this showed that the southern continents had not always been in their present position.

▲ A frond of the Permian seed fern *Glossopteris* from the Transvaal, South Africa.

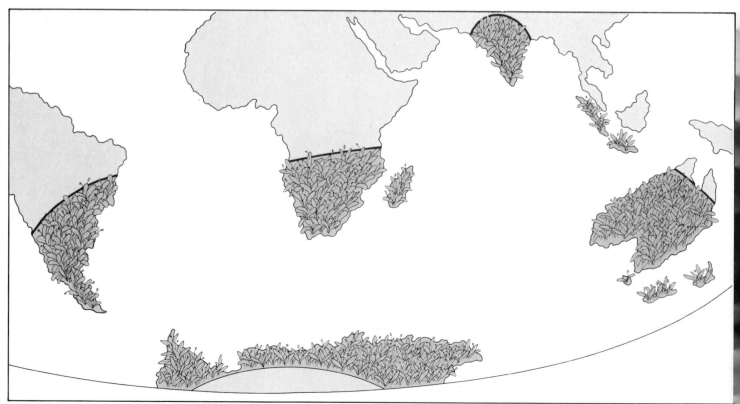

▲ The distribution of fossil *Glossopteris* flora today. This is difficult to understand unless the southern continents were once joined together in the great Palaeozoic southern continent of Gondwanaland as shown in the second drawing.

iron minerals in contrast to the lighter granitic 'sial' of silica and alumina minerals. This meant that the lighter granitic continents could never have sunk beneath the seas. Only shallow marine water deposits have ever been found on the present-day continents, suggesting that continents and oceans were fundamentally distinct. The ocean basins therefore seemed to have been permanent features of the Earth.

This was an important argument when it came to explaining the distribution of animals and plants. It had been customary for palaeontologists to postulate the existence of land bridges to account for the similarity of animals on different continents, but, as Wegener stressed, there is no sign of such bridges and, had they existed, there is no way in

which they could have sunk down into the denser sima. If it is accepted that similarities in animals and plants result from their genetic material, the presence of the same types on different continents cannot be explained unless the separate land masses were once in contact. The existence of the same plants, the *Glossopteris* flora, in South America, Africa and India suggested a direct connection between these three regions. Wegener also compared the geological structures along the eastern coast of Brazil with those on the south-western coast of Africa. 'It is just as if we were to refit the torn pieces of a newspaper by matching their edges and then checking whether the lines of print run smoothly across. If they do, there is nothing left but to conclude that the pieces were in fact joined in

this way.' Wegener's most convincing evidence was drawn from his knowledge of climates, especially those of the past. He pointed to the evidence of an ice age during the Permo-Carboniferous, in particular the presence of tillites or boulder clays deposited from glaciers and striated rocks, caused by glacial action. These were found in India, Australia, Africa and South America. This distribution of ice-age phenomena was incomprehensible if the continents had always been situated as they are today. Similarly, the Upper Carboniferous coals, which represented the equatorial swamps of the humid tropics, occurred in North America, western Europe and in northern China. Immediately to the north of the coal deposits were evaporite rocks formed under dry, hot, desert conditions. So it seemed as though today's hot southern regions of the Earth had been covered in ice while the northern zone had enjoyed a tropical climate. The response of geologists was to suggest that coal was not necessarily evidence of equatorial conditions and, similarly, that many of the supposed tillites had probably been misinterpreted. However, Wegener suggested that about 275 million years ago all the continents had been united in a single land mass which he named Pangaea. The South Pole had been close to the junction of Africa, Antarctica, India and Australia and the equator had run through North America, western Europe and Siberia to China. This new interpretation overcame the

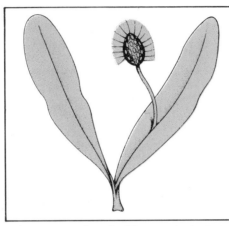

▲ A reconstruction of a *Glossopteris* plant which was deciduous.

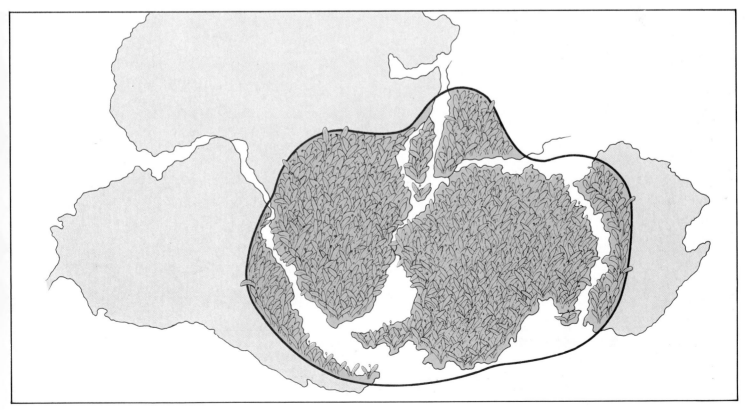

problem of glacial deposits and coal occurring in the 'wrong' latitudes.

The brilliant concept of large horizontal displacements of the continents, later known as 'continental drift', immediately accounted for many observations that had puzzled geologists for a long time. Yet Wegener's synthesis was firmly rejected by the geological community so that he had difficulty in obtaining a university teaching post in his own country, and eventually found one at the University of Graz in Austria. Wegener was very much a generalist at a time when scientists made a virtue of being specialists. Although he was disparaged in his lifetime, today Wegener is accepted as a major figure in the Earth sciences.

Wegener was unable to explain the mechanism by which the continents could have moved, although he made some tentative suggestions which the physicists were able to demolish with ease. It was stressed that the so-called continental fit was not accurate and indeed a number of the claims he had made could not be backed by evidence. The physicists noted that there was geological evidence in support of Wegener's theory, but that it was suspect because there was no physical evidence to substantiate it.

Fortunately Wegener's ideas did occasionally fall on fertile ground. The geologist Alexander L. du Toit (1878–1948) of South Africa took up the torch. His first contribution was in 1927, three years before Wegener's death, but his most influential book was *Our Wandering Continents* published in 1937. He dealt with the question of the fit of Africa and South America by pointing out that it was not the present-day coastline that was significant, but rather the edges of the continental shelves. But du Toit's great contribution stemmed from his detailed knowledge of the geology of the southern continents. He was able to reassemble the former great Palaeozoic southern continent of Gondwanaland. He demonstrated that there was an ancient structure, which he named the 'Samfrau geosyncline', that extended from Argentina through southern Africa into eastern Australia, showing that they had all been linked together. Although the geological community remained dismissive of his views, du Toit's book had a wide influence.

There was one other major figure in the geological world who had rallied to the support of Wegener – the British geologist Arthur Holmes (1890–1965). In 1929 he postulated that the mechanism for continental drift was a series of convection cells within the mantle beneath the Earth's crust. His most powerful and significant support for Wegener's theory came with the appearance of his text-book *Principles of Physical Geology*, published in 1944. Holmes devoted the last chapter of his book to continental drift, stating that the geological evidence was overwhelming, the only serious objection being the difficulty of finding a drift mechanism. He suggested that the continents, like rafts, were carried

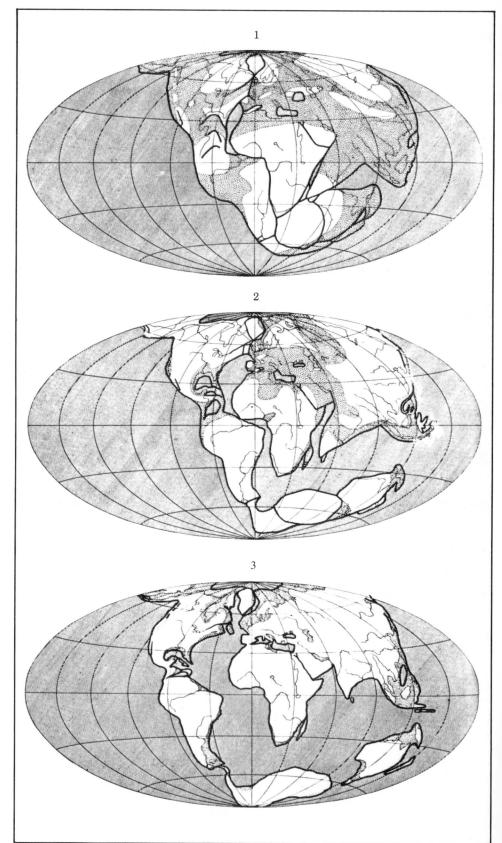

▲ The illustrations are from Alfred Wegener's book *The Origin of Continents and Oceans*. They show what he considered to be the positions of the oceans and continents in: 1 the Upper Carboniferous – all the continents were united in the super-continent of Pangaea; 2 the Eocene – the continents had begun to move apart; 3 the beginning of the Quaternary – the continents were not far from their present-day positions. (The dotted areas represent shallow seas.)

by convection currents within the mantle generated by heat rising from the core. This text-book stands as a landmark as it is one of the most readable books on geology ever written. The generations brought up on this text have been introduced to the concept of continental drift in its most persuasive context.

By the 1950s and 1960s, owing to Holmes, the geological community was receptive to Wegener's ideas. The evidence that finally established that the continents had moved relative to one another came from detailed studies of the ocean floors. These began just before the outbreak of the Second World War when the Dutch geophysicist, F. A. Vening Meinesz, began exploring the continental shelf using seismic methods and measuring gravity with instruments in submarines. The sediments on the ocean floors were never more than a kilometre in thickness, which is exceptionally thin. There were no signs of any continental rocks anywhere in the oceans and all oceanic rocks were basaltic. Furthermore, the boundary between the crust and the underlying mantle was at a depth of about 10 km, whereas beneath the continents it was some 30 km deep. Vening Meinesz was able to prove conclusively that the ocean basins could never have been formed by the sinking of continental masses. By far the most surprising discovery he made was that there were no ocean floors older than the Cretaceous period. Most of them were much younger even than this. While he could demonstrate that there were fundamental differences between the continents and the ocean floors, it was impossible to explain why, on land, there were rocks thousands of millions of years old, whereas the oceans were floored by rocks that were less than one hundred million years old.

The interest aroused by the new data also led to some of the old evidence being re-investigated. The apparent fit of the continents that had been a subject of dispute was resolved completely by the English geophysicist Sir Edward Bullard. It had been emphasized that Wegener had crudely overstated the case for the continents fitting together, but Bullard and his colleagues, using a sophisticated geometrical technique and computers, produced an astonishingly accurate match of the continental shelves. It was accepted, henceforth, that the two sides of the Atlantic matched. Clearly there is no way this can be explained apart from postulating that the two continents had been united and had subsequently become separated and drifted apart. Even fossil evidence which had been collected showed that the successions of fossil reptiles in South America and southern Africa were beginning to match in great detail; the same major reptile faunas were even discovered in Antarctica – for example, the hippopotamus-like *Lystrosaurus* was known also from southern Africa and India. As more evidence came to light, it all lent further support to Wegener's theories.

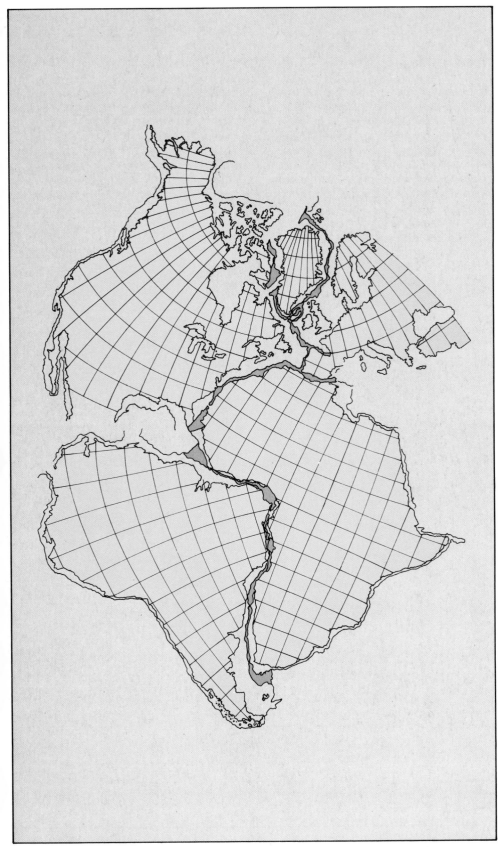

▲ The maps produced by E. C. Bullard, J. E. Everett and A. G. Smith using sophisticated geometry and computers, show an accurate fit of the continental shelves of the continents bordering the Atlantic.

THE ANCIENT POLES

The examination of magnetic rocks showed that the ancient poles seemed to have changed position many times. The only rational explanation for this was continental drift.

The Earth has long been known to act as a magnet. William Gilbert, physician to Queen Elizabeth I of England, wrote on the Earth's magnetism as long ago as 1600. When iron filings are sprinkled around a bar magnet, they take up positions which illustrate the lines of force generated by the magnet. In the same way, atoms in an object affected by magnetism will align themselves along the lines of magnetic force and point to one of the magnetic poles. Normally an arrangement of atoms is entirely random; however, in some minerals the arrangement of the crystals facilitates the magnetic alignment of the atoms. This happens with the appropriately named iron oxide, magnetite, Fe_2O_4.

The Earth is thought to have an iron-nickel core, which, as the Earth rotates, induces magnetic forces. These extend far beyond the surface of the Earth to form the magnetosphere. In the magnetosphere, the Van Allen belts of magnetized particles give the same type of pattern as do iron filings. A magnet, if carefully suspended, will point to the Earth's magnetic pole, and therefore can be used to give direction. This is the principle of the compass.

It has been known since the time of the early Chinese that certain rocks, called 'lodestones', naturally act like magnets, so that if a suitably shaped rock is suspended freely, it will point north. This is because it contains magnetite, which is among the most common minerals in igneous rocks. When the igneous rock was formed, the grains of magnetite crystallized out of the molten magma and oriented their polarity in the Earth's magnetic field. Once this has taken place, the magnetization is preserved.

One of the major tools in geophysical prospecting is recording the magnetism in rocks where there are mineral deposits like magnetite. It was as a by-product of studying rock magnetism in the course of oil and mineral exploration that it was discovered that minerals preserved their original magnetism. It was possible to record the magnetism at different geological ages and when this was done it became clear that the Earth's magnetic poles appeared to be in different places at different times in the past. The plotting of the apparent position of the magnetic North Pole through geological time showed a wandering path. A concept of polar

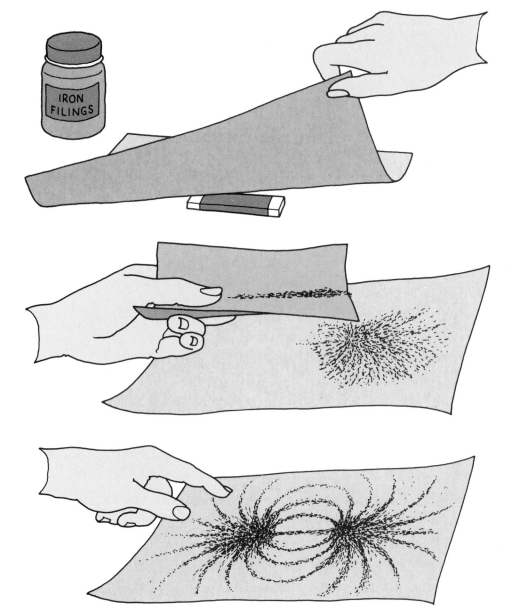

▲ The Earth's magnetosphere or magnetic field. This extends for thousands of kilometres into space and its shape is affected by the solar wind, a stream of atomic particles emitted by the Sun.

Magnetism experiment
◀ Place a piece of paper over a bar magnet, then sprinkle iron filings on top of the paper. The filings will align with the lines of force generated by the magnet. Atoms in some igneous rocks behave in the same way and align with the Earth's poles.

wandering was put forward, which suggested that over millions of years the axis of rotation of the Earth must have been changing so that, relative to the continents, it appeared as if the position of the geographical poles had moved. Alternatively, the poles could have remained in much the same place while the continents had moved.

There was a simple test to decide which of these theories was correct and that was to examine the polar wandering as recorded in different continents. Once this was done, it was discovered that North America and Europe had different tracks of polar wandering. Now there is no possibility that the Earth had two separate North Poles at the same time, and in fact the polar wandering tracks converged until they met up at what is now the magnetic pole. The only explanation of these observations was that the relative positions of the two continents had been different in the past. Subsequent work showed that the polar wanderings of all the continents were independent, which proved that

none of them was positioned as it is at the present day.

If the southern continents are united as the single continent of Gondwanaland, the puzzle of these divergencies of polar wandering can be overcome. Continental drift was thus established as a historical fact, but produced in its wake a series of seemingly intractable problems. The long-standing question still remained: what mechanism could have caused continental drift? It was accepted that the continents were now in different positions relative to one another and that the light continental material floated on the dense basaltic oceanic crust. But the continents were solid rock between 25 and 75 km thick and there was no hint of what could have caused them to move.

The evidence from palaeomagnetism coincided with the palaeoclimatic evidence originally presented by Wegener (see pp 180–1) and was an eloquent confirmation of his original hypothesis of continental drift. However, palaeomagnetism could

only provide evidence of the latitudes at which the continents had been positioned. There was no way of determining the longitude. At any given longitude the pole would always be in the same position.

Many of the solutions to the problems regarding the movements of continents came from surveys of the ocean floor in which magnetometers were towed behind survey ships to record the Earth's magnetic field. To everyone's surprise, instead of the expected rather flat monotonous profile, the surveys produced a striking pattern of alternating high and low magnetism running parallel to the mid-oceanic ridges. Already in the 1950s it had been discovered that the Earth's magnetic field periodically reversed so that the compass instead of pointing north would point south. The changeover seemed to take about 5,000 years and then remained more or less constant for 10,000 to 150,000 years. This curious observation was to lead to the discovery of the mechanism of continental drift.

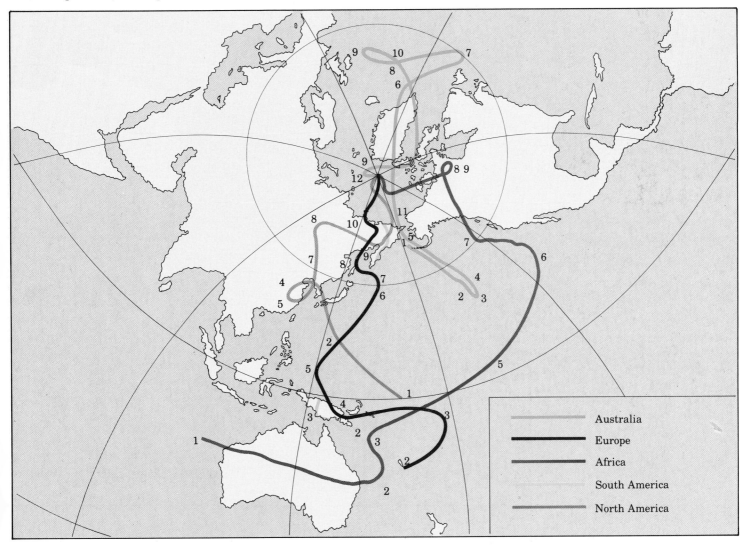

▲ The map shows independent polar wandering curves for various continents over geological time based on palaeomagnetic studies. This is evidence that the positions of the continents have altered.

1 Precambrian 2 Cambrian 3 Ordovician
4 Silurian 5 Devonian 6 Carboniferous
7 Permian 8 Triassic 9 Jurassic
10 Cretaceous 11 Tertiary 12 Quaternary

SEA-FLOOR SPREADING

The concept of the sea-floor spreading by generating new material explained the mechanism of continental drift.

During the Second World War sophisticated echo-sounding devices were developed for the detection of submarines. Using these, it was discovered that the topography of the sea-bed had a regularity that was quite distinct from that of the land. Huge mountain chains rose 3 to 5 km above the floor in the middle of all the oceans. These ranges, which ran north-south, were up to 1,500 km wide and running along the mid-crest was a rift-like valley up to 3 km deep and some 60 km wide. This central valley was similar to the great rift valleys of East Africa which were considered to be tensional structures, where the crust was being pulled apart. These mid-oceanic ridges, as for example the mid-Atlantic ridge, ran parallel to the edges of the continental shelves of the opposing coastlines. When these curved, the mid-oceanic ridge followed, not by bending round, but by fracturing at right angles – that is, east-west – to the crests; these fractures are known as 'transform faults'. Parallel to the mid- oceanic ridges were innumerable ridges and valleys.

The topography and the apparent tensional structures at the crest of the ridges seemed to be present in all the world's oceans which together made up some 40,000 km of mid-ocean ridges. In 1960 the American oceanographer Harry Hess postulated that the mid-oceanic ridges were the sites of new ocean crust being generated: this was his theory of sea-floor spreading.

In 1963 the British scientists Frederick J. Vine and Drummond H. Matthews drew together the various lines of evidence from a detailed study of the mid-Atlantic ridge, south of Iceland. Detailed surveys of the magnetism of the ocean floor had shown that magnetic variations form a simple pattern of long strips, varying in width. This pattern was formed as the igneous rocks of the ocean floors were magnetized alternately in opposite directions, according to the changing polarity of the Earth's magnetic field (see pp 184–5). The parallel magnetic strips were in fact exactly symmetrical on either side of the mid-oceanic ridges. The different widths of the strips also matched the different proportions of the alternating zones of normal and reversed polarity that had already been worked out using both fossil evidence and radioactive dating. Taking South America and Africa as an example, Vine and Matthews proposed that the two continents were once joined, but had then fractured. As they came apart, molten material welled up forming a ridge to fill the space. When this cooled to below 450°C, it became magnetized in the direction of the Earth's magnetic poles at the time. The continents would have been pushed further apart as new molten rock welled up through the ridge already formed. This in turn was magnetized, recording any changes in polarity. As more molten rock rose and became magnetized, the older rocks were pushed further and further away on either side of the ridge, carrying the continents with them and preserving

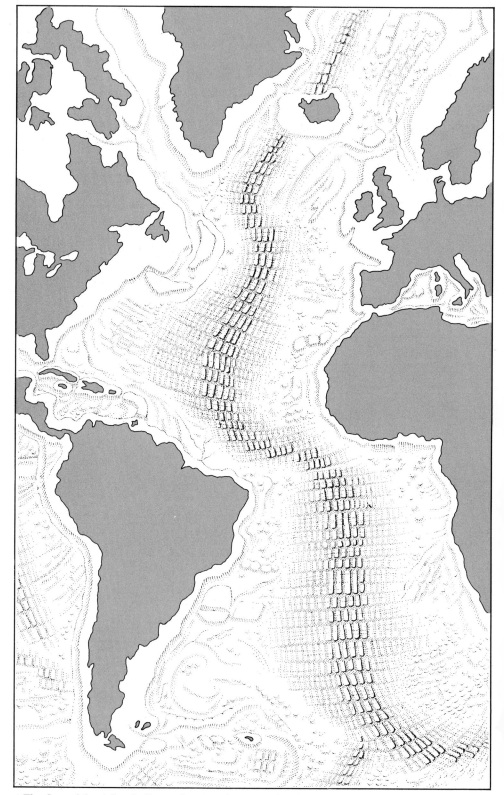

▲The floor of the Atlantic Ocean showing the mid-oceanic ridge with transform faults running at right angles to it. It is through such ridges that new oceanic crust is generated.

the record of polarity reversals. This meant that further away from the central ridge, the ocean floor became progressively older. The mechanism also explained the identical sequence of magnetic strips on either side.

This new concept of continents moving simply because new ocean floor was being generated between them overcame the difficulty of discovering a mechanism by which the continents could move. Their role in the process suddenly became entirely passive. This supremely elegant theory, which backed Hess's original speculation by providing some firm evidence, accelerated research. Within a few years the same pattern of magnetic strips was recognized throughout the world. Deep-sea drilling for samples of the sediment immediately overlying the igneous ocean floors has enabled the age of the strips to be checked from the microscopic fossils they contain.

Again taking South America and Africa as an example, it is possible to work out exactly how wide the southern Atlantic Ocean has been at specific times in the past by the age and distance from the central ridge of the magnetic strips. From such studies it has become possible to measure the rate at which the continents are moving apart and the oceans spreading. Overall the rate varies between 2 and 14 cm per year. It is known that Africa and South America were united in the Cretaceous period and only began to drift apart towards the end of the Cretaceous.

The deep-sea drilling, which confirmed the theory of Vine and Matthews, had further implications. The exact dating of the rocks led in 1976 to the confirmation of Milankovitch's astronomical theory of climatic cycles and, consequently, the causes of the ice ages (see pp 166–7). As far as evolution is concerned, the sequences in the rocks of microscopic organisms over some 50 million years have demonstrated that the changes that took place were gradual and affected entire populations. Where the fossil record is complete, phyletic gradualism is seen to have occurred. Where there are gaps, one can postulate jumps or catastrophes, but they can also be seen as due to adverse conditions of preservation or succeeding migrations and are not necessarily a reflection of a different kind of process. There is however one feature that seems to be a regular occurrence. Many microscopic organisms suddenly die out at exactly the point of magnetic reversal. Some organisms appear to be unaffected, but certain species disappear. This happens sufficiently regularly for it to be unlikely to be mere coincidence, but no one knows the reason for it.

Sea-floor spreading is caused by convection currents in the mantle beneath the crust. These convection cells, which bring heat generated by radioactivity deep within the Earth, may well cause magnetic reversals when the convection current changes direction. The Earth is protected from the direct effects of the radiation from outer space by the magnetosphere and in particular the Van Allen belts. But during a change in polarity, it is thought that the magnetosphere allows the radiation through, although there is no knowledge of exactly how much, or of what effect this has on the Earth's surface. At the moment this is an area of speculation and any theories will have to await either confirmation or refutation until the next reversal takes place.

▲Pillow lava at the mid-Atlantic ridge. When lavas are extruded under water at mid-ocean ridges they often form into distorted globular shapes.

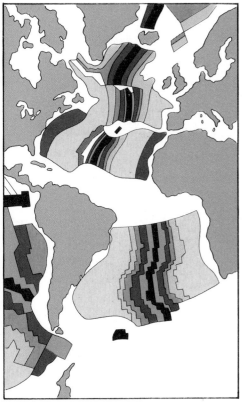

▲Different ages of the rocks on the floor of the Atlantic Ocean (the unmarked areas do not have enough data). The rocks become progressively older on either side of the mid-oceanic ridge, which is proof of sea-floor spreading.

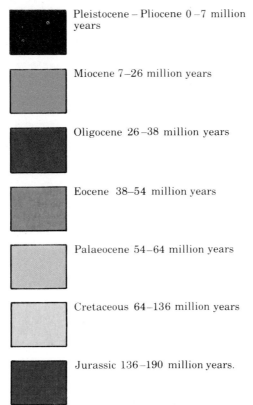

Pleistocene – Pliocene 0 –7 million years

Miocene 7–26 million years

Oligocene 26–38 million years

Eocene 38–54 million years

Palaeocene 54–64 million years

Cretaceous 64–136 million years

Jurassic 136–190 million years.

SUBDUCTION ZONES

**As the ocean crust is generated, it is also destroyed
by being drawn back into the Earth's mantle through subduction zones.
This process causes deep-seated earthquakes.**

In the early 1960s it was accepted from the evidence of palaeomagnetism that the continents had moved. The evidence of sea-floor spreading revealed the means by which this had occurred. This still left a problem. From the evidence available, it was clear that new oceanic crust was being produced at a fairly rapid rate, and so it was necessary to postulate either an expanding Earth or that oceanic crust was being destroyed at the same rate as it was being generated. It has been calculated that if the Earth had been smaller 200 million years ago, animals would have weighed four times as much as they do now for the same volume; and in the case of the large dinosaurs, the strength of their bones would have been quite incapable of

supporting them. This kind of evidence has led scientists away from the expanding Earth idea. Most geologists today recognize that the dimensions of the Earth can be considered constant. The accepted view is that the ocean crust is being continually destroyed. The convection current, which causes the ocean crust to move like a conveyor belt, begins at the mid-oceanic ridges, where it rises up from the mantle, runs away and then dips down back into the mantle in a giant convection cell – essentially the system of convection cells postulated by Arthur Holmes in 1929 (see pp 182–3).

In his first outline of sea-floor spreading in 1960 Hess had suggested that the deep-sea trenches may have been the site at which the oceanic crust

was drawn back down into the mantle. The deep-sea trenches are furrows on the bed of the oceans some 9–11 km deep, that is, twice the normal ocean depths; they are generally 250–300 km wide and form an arc. To the landward side there frequently occur island arcs which are of volcanic origin and run parallel to the deep-sea trenches. In the 1930s Vening Meinesz had recorded gravitational anomalies beneath island arcs. Island arc regions were known to be active earthquake zones, but what was significant was that the earthquakes took place along a plane (called the Benioff zone) inclined at about 45° dipping down from the region of the trenches towards the island arcs and extending downwards for 700 km. This information led Hess

to conclude that it was in these regions that the oceanic crust was thrust downwards, that here were subduction zones where the oceanic crust was destroyed. It was the friction along this plane which resulted in the existence of earthquakes.

As the result of the nuclear test ban treaty, the United States set up a world-wide network of stations to record earthquakes, as a prerequisite to being able to locate any underground nuclear explosions which might be claimed to be earthquakes. It was the natural suspicion between the great powers that led to some 30,000 earthquake shocks being monitored between 1961 and 1967. This mass of information provided key support for Hess's theory. It became clear that volcanoes and earthquake activity were restricted to very narrow zones. They seemed to be concentrated along mid-oceanic ridges, deep-sea trenches and island arcs. They also occurred along transform faults joining mid-oceanic ridges to island arcs. It was equally evident that most regions of the world

did not suffer from either volcanoes or earthquakes. In fact the surface of the Earth could be divided into a number of major regions which were bounded by earthquake and volcanic zones. These regions of the Earth's crust were called plates, and were composed either entirely of oceanic crust or of partly continental and partly oceanic crust. At the plate margins where the crust was being generated, there was often volcanic activity as magma welled up from the mantle; there were also innumerable earthquakes, although these tended to be rather shallow. At the margins where the crust was being subducted to be reincorporated into the mantle, earthquakes were generated down the Benioff zone, and it is in this zone only that the deep-seated earthquakes originated. The existence of volcanic island arcs beyond the deep-sea trenches but parallel to them is due to magma rising from the edge of the crust as it sinks sufficiently deep to be remelted, disturbing the mantle. This is why volcanoes are associated with subduction zones not at their

surface expression in the form of deep-sea trenches, but where the line of friction down the plane of the Benioff zone reaches the region at about a 700 km depth, where the mobilization of magma takes place causing volcanic activity.

With the recognition of subduction zones it was now possible for the first time to begin to understand the fundamental dynamics of the planet Earth. Major convection cells within the mantle draw magma to the surface, where it solidifies and is moved away from the source in a never-ending stream until it is dragged back down by the same convection cell, as the currents deep in the mantle turn over. The continents float on the surface of the denser basaltic crust and are never dragged back down to be recycled, rather like soap-suds in the bath which remain floating as the water beneath them runs away. The oceanic crust is being continually recycled so that there can never be any ancient oceanic crust preserved; but the continental crust remains floating.

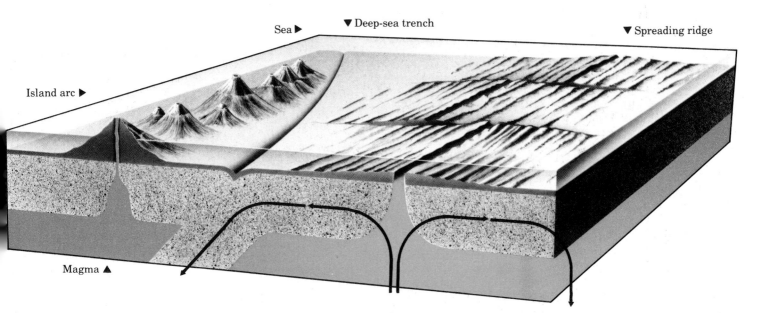

Sea ▶   ▼ Deep-sea trench   ▼ Spreading ridge

Island arc ▶

Magma ▲

▲ The convection currents, which make the ocean crust move like a conveyor belt, begin at the mid-oceanic ridges and dip back into the Earth's mantle at the subduction zones.

▶ Map of the world showing earthquake epicentres based on a world-wide seismic survey, 1961–7. This revealed that earthquake and volcanic activity were restricted to very narrow zones.

◀ Island arc near the Californian coast. Island arcs run parallel to deep-sea trenches and are caused by volcanic activity.

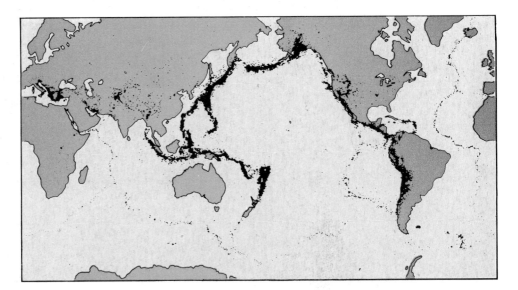

PLATE TECTONICS

The recent theory of plate tectonics has shown that mountains were formed by the collision of different plates of the Earth's crust.

The unifying theory of evolution transformed the biological sciences in the nineteenth century. Just over a hundred years later the geological sciences witnessed a similar intellectual revolution with the unifying theory of plate tectonics. It had long been known that there were a number of mountain-building episodes in the history of the Earth. Even the ancient Greeks had accepted that sediments from the sea-bed had been raised up to form the highest mountains, but it was not until the 1960s that any sensible answer could be provided to explain the mechanism which caused the formation of mountains.

The Earth's crust consists of a series of interlocking rigid plates, which are in a state of continuous movement caused by convection cells in the underlying mantle. The boundaries of the plates are marked by volcanic activity and earthquakes, which are both restricted to the regions where the ocean crust is being generated and those where it is being destroyed. It is at the plate margins furthest away from the generative mid-oceanic ridges that the more dramatic events take place. This is where two plates, instead of moving away from one another as a consequence of the formation of new oceanic crust, are in the process of colliding. There are three types of collision that can occur: oceanic crust can collide with oceanic crust, oceanic and continental margins can meet, or there can be a direct collision between continental crusts; each will result in mountain building.

When two plates of oceanic crust meet one another, the older plate, by virtue of its age, will be denser than the younger and will be more readily forced down beneath it. The expression of this on the sea-floor is a deep-sea trench. By the time the sinking edge of the plate has reached such a depth that it is becoming reincorporated into the mantle, the heat being generated will lead to the mobilization of magma. Volcanoes develop on the sea-floor, building up sufficient solidified magma to rise above sea-level and form an island arc, parallel to the deep-sea trench and ahead of the line of the

subduction zone (see pp 188–9). The island arcs around the western Pacific are classic examples of the results of the collisions of oceanic plates.

Where the margin of one plate carries continental crust and comes into contact with oceanic crust, the oceanic crust is simply driven beneath it. The continental margin is lifted up, and huge elongated granitic batholiths are formed from magma generated by friction and melting at depth. The molten magma melts the rocks of the

continental crust, changing their composition, hence the granitic nature of the intrusions. The Andes of South America are an example of this type of collision, where there is evidence of considerable volcanic activity together with the emplacement of vast quantities of granites. The Andes include some of the world's highest mountains which are the result of direct vertical movement with little evidence of horizontal displacement.

Finally, when two masses of

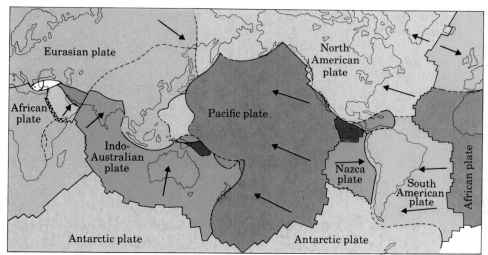

▲ The map shows how the Earth's crust can be divided up into a series of plates. Each plate is usually composed of some ocean floor as well as continental mass. The plates are moving in the directions shown by the arrows; this movement is caused by the action of convection cells in the underlying mantle.

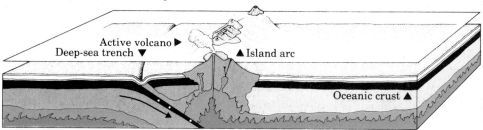

Mountain building
▲ Mountain building results from collisions between the Earth's plates. 1 When two oceanic plates meet, the older plate will sink beneath the younger causing a deep-sea trench. The sinking plate becomes reincorporated into the mantle; this generates heat and magma is mobilized resulting in volcanic activity which forms an island arc.

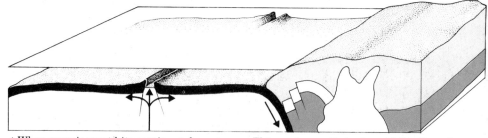

▲ When oceanic crust hits continental crust, the oceanic crust sinks beneath the lighter continental material, raising the continent. This again results in the mobilization of magma so that longitudinal batholiths or granite intrusions form.

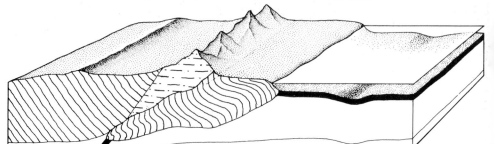

continental crust are in collision, one continental mass is thrust into another and this results in intense folding and faulting. Igneous intrusions invade the roots of the mountain chain and rock strata are folded over on themselves in complex structures. These are seen in the Himalayas, where India was overridden into the southern margin of Asia 15 million years ago, and in the Alps, where the African plate crashed into Europe 26 million years ago. Indeed some parts of the present-day Alps comprise slivers of African rocks. The volcanic activity in southern Europe and Asia Minor together with frequent earthquakes are a measure of the continuing movement of the Eurasian and African plates, as the African plate is still being driven northwards. The deeper, more basaltic layers have plunged beneath the Eurasian plate.

The continents contain rocks of all geological ages, even some as old as 3,800 million years old, because the continental masses are never recycled in the same way as oceanic crust. In fact in the process of plate tectonics, newer rocks are added onto the continental margins.

In the light of plate tectonics, it becomes valuable to re-examine such topics as the former distribution of animals and plants. The formation of an ice-cap in the Antarctic towards the middle of the Tertiary seems to have been due to that continent moving into the position of the southern magnetic pole. The resultant climatic change and the spread of grasslands had a profound effect on the evolution of the mammals. The movement of the African plate, so that it collided with the Eurasian plate in the Miocene, enabled the grassland-dwelling primates, the antecedents of man, to spread into temperate latitudes. The pattern of extinctions throughout the fossil record seems to be related to global changes of sea-level, and these in their turn are the direct result of the generation of oceanic crust. The distribution of land animals can be properly understood once the sequence of the break-up of the single super-continent Pangaea 180 million years ago, and of the southern continent Gondwanaland 50 million years later, has been carefully documented.

It is now becoming possible to trace the history of continental masses prior to the formation of Pangaea which took place in the Permian. The great Caledonian mountain-building period during the late Silurian and Devonian is now known to have been the result of the closure of the Iapetus Ocean when what is now eastern America and western Europe collided to form the Old Red Continent; the line of suture or collision was marked by a huge mountain chain, the Caledonides. As the oceanic crust dived beneath the margin of its adjacent plate, the continental mass would have been carried on the moving belt of oceanic crust, and the Iapetus Ocean, the proto-Atlantic, would have gradually closed up and the two continental masses collided to form the Caledonides.

The sequence of major climatic changes recorded in the fossil record of Europe and North America simply documents the movements of plates which have carried great continental masses across the climatic zones of the Earth's surface. Wegener pointed the way with his perceptive studies of the distribution of rock types which were climatic indicators.

Together the theories of evolution and plate tectonics have provided man with fundamental insights into his place on Earth.

▲Mt Huayna Potosi, Cordillera Real, Bolivia, which forms part of the Andes. The Andes are the result of mountain building caused by oceanic crust and continental mass colliding.

◀When two continental masses collide, intense folding and faulting of the rocks results. Magma is mobilized and igneous intrusions form at the roots of the mountains.

▲Ladakh in the Himalayas, which were formed when the Indian plate crashed into the Eurasian plate. The strata on the mountain are nearly vertical, which is evidence of intense folding.

APPENDIX

TIME

Until the beginning of the nineteenth century, the Earth was generally considered to be some 6,000 years old. Bishop Ussher had calculated that the Creation took place in 4004 BC on the authority of Biblical chronology. But by the beginning of the early 1800s, the geologists and palaeontologists recognized that the evidence of rocks and fossils precluded a literal interpretation of the Bible.

The recognition of thick sequences of rocks characterized by different types of fossils led to an acceptance of the antiquity of the Earth and the existence of an antediluvian world. The father of stratigraphy, British canal engineer William Smith (1769–1839), produced in 1815 the very first geological maps, in which he showed the distribution of rocks in England. These were followed in 1816 by the book *Strata Identified by Organized Fossils*. William Smith was the first to recognize that rocks could be identified from the series of fossils they contain. Furthermore,

the fossils allowed sequences of rocks to be correlated from country to country and even from continent to continent.

The rock strata divide into a number of geological periods, which in the main are named after the region in which they were first scientifically studied. Two pioneering British geologists in the nineteenth century were Adam Sedgwick of Cambridge University and Roderick Impey Murchison who studied the oldest British fossil-bearing rocks in Wales. Sedgwick named the Cambrian after the Roman name for Wales, and Murchison the Silurian after the Silures, a Celtic tribe of the region. Sedgwick and Murchison worked together, but their friendship foundered over a sequence of rocks which Murchison insisted were Lower Silurian and Sedgwick, Upper Cambrian. Such was the vehemence with which they held their views that they became bitter enemies to the end of their days. After their deaths in the early 1870s, Professor Lapworth of Birmingham University in 1879 proposed a new geological period for the disputed sequence, the Ordovician, named after the ancient Ordovices tribe of central Wales.

It was Charles Lyell who subdivided the Tertiary or Cenozoic (from the Greek *kainos* recent; *zoe* life) era 'beginning 64 million years ago' on quite a different basis. As most of the Cenozoic fossils are gastropods and bivalves, the geological periods were divided and named according to the proportion of living species of molluscs in existence at the time: the Eocene is the 'dawning' (from the Greek *eos*, dawn) of the recent, the Oligocene means 'few', the Miocene means 'smaller' in number, the Pliocene 'greater' in number, and finally the Pleistocene means most 'numerous'. The division into the major eras of Primary, Secondary and Tertiary had already been established by Giovanni Arduino in the eighteenth century. The names Palaeozoic or 'ancient life', Mesozoic or 'middle life' and Cenozoic or 'recent life' were introduced by John Phillips in 1841.

An early observation was that certain types of fossil were characteristic of particular periods of the Earth's history. Hence the Cambrian, Ordovician and Silurian seemed to be dominated throughout the world by trilobites. During the Ordovician and Silurian, a group of

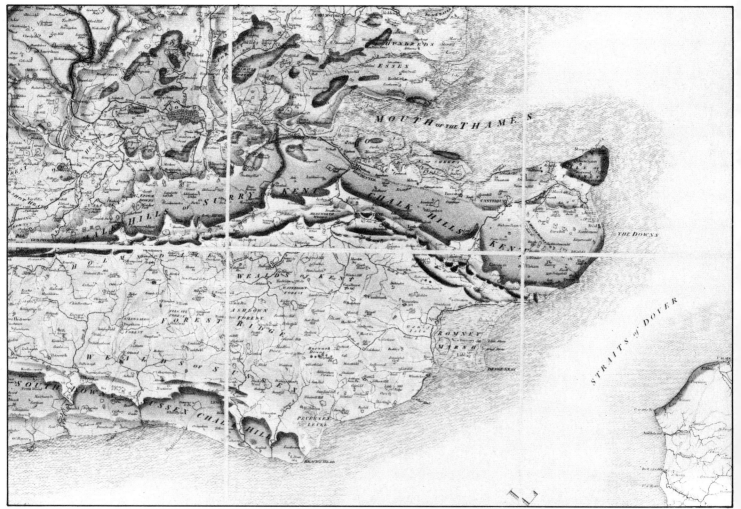

▲ Part of William Smith's *A Delineation of the Strata of England and Wales, with part of Scotland* (1815), which was the first large-scale geological map. The map allowed particular strata such as the Chalk to be traced over hundreds of kilometres; Smith's work established the basis for a detailed scientific study of the succession of rock strata.

▲The Grand Canyon of the Colorado River, Arizona, USA, is the classic example of a sequence of horizontal strata which illustrates the law of superposition (which states that if one series of rocks lies above another, then the upper series was formed later). The 1.5 km deep gorge was formed when the land was uplifted and the existing rivers cut down to expose the succession of strata.

planktonic colonial organisms, the graptolites (so-called because they looked like writing on stone), occur in most marine rocks throughout the world. From Devonian times up to the end of the Cretaceous period, the characteristic fossils are ammonoids, and during the Tertiary, the sedimentary rocks abound with bivalve and gastropod molluscs. For fossils to be of use in relative dating, using rocks over large areas, they need to have a wide distribution, hence free-swimming planktonic organisms are ideal. Furthermore they need to be forms which undergo rapid changes; in this respect trilobites, graptolites and ammonites make excellent examples for dating.

Geologists were interested in finding out the real age of the sedimentary rocks. Many different attempts were made to estimate this using Lyell's criteria of examining the processes taking place at the present day. On the basis of the changes in the mollusc faunas Lyell himself suggested an age for the Tertiary era of some 80 million years and for the Cambrian period of 240 million. Another method was to calculate the inferred rate of deposition of sediments, and from the thickness of the strata it would be possible to assess the minimum age. This method gave a range of a 100 to 600 million years for the beginning of the Cambrian. In virtually all cases the suggested timespans turned out to have been grossly underestimated. One novel method was utilized by polymath John Joly of Dublin, Ireland: from the amount of salts reaching the world's oceans, he calculated that the present-day amounts must have accumulated over 80 to 100 million years. Darwin, in *The Origin of Species*, unlike his contemporaries came up with figures now known to be considerably more than the real timespans (see p 11).

While the geologists were proposing ages in terms of hundreds of millions of years, as this is what their empirical observations demanded, the physicist William Thompson, later Lord Kelvin (1824–1907), applied his mathematical acumen to the question. Kelvin was able to 'prove' from his calculations that the Earth would have taken between 20 and 40 million years to cool to its present condition from an original molten state. The geologists insisted that the geological column established a far greater time-scale than that allowed by Kelvin, but they were unable to counter his mathematical arguments. At the end of the nineteenth century radioactivity was discovered and in 1903 it was established that radioactivity generated

Era	Period	Epoch	years ago
Cenozoic	Quaternary	Holocene	10,000
		Pleistocene	2 million
	Tertiary	Pliocene	7 million
		Miocene	26 million
		Oligocene	38 million
		Eocene	54 million
		Palaeocene	64 million
Mesozoic	Cretaceous		136 million
	Jurassic		190 million
	Triassic		225 million
Palaeozoic	Permian		280 million
	Carboniferous		345 million
	Devonian		410 million
	Silurian		440 million
	Ordovician		530 million
	Cambrian		570 million

▲ The column shows how geological time is divided into different periods.

▲ The Sphinx at Giza, Egypt is a natural Eocene rock formation showing layers of horizontal strata. The basic shape was formed by the wind and it was carved *in situ*.

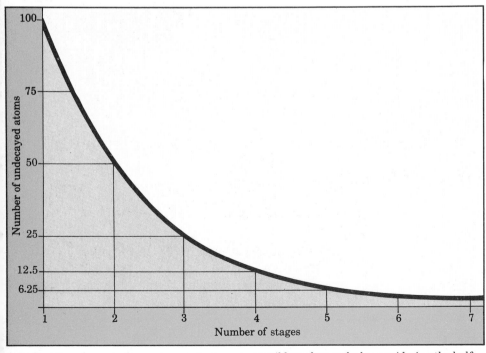

The time-scale chart on the left:

Hundreds of million years ago	
0	Age of mammals — Age of dinosaurs — First land vertebrates
5	First skeleton (good fossil record starts here) — First multicellular animals (Ediacara)
10	
15	First eukaryotes
20	
25	
30	
35	First fossils (bacteria, banded ironstone formation, stromatolites)
40	Oldest rocks
45	Moon/meteorites

▲ The time-scale of the history of the Earth illustrates the vast length of time that elapsed before an abundant fossil record existed.

Radioactive isotope	Half life	Daughter isotope
Uranium 238	4,510 m.y.	Lead 206 (only over 20 m.y.)
Uranium 235	713 m.y.	Lead 207 (only over 20 m.y.)
Thorium 232	13,900 m.y.	Lead 208 (only over 50 m.y.)
Potassium 40	1,300 m.y.	Argon 40 (only over 100,000 years)
Rubidium 87	47,000 m.y.	Strontium 87 (only over 10 m.y.)
Carbon 14	5,570 years	Nitrogen 14 (not beyond 70,000 years)
Strontium 90	27 years	No value in geology
Radon 220	52 seconds	No value in geology

▲ Table of radioactive isotopes, their half life and daughter isotopes. Some radioactive isotopes are more useful for measuring ages than others.

▲ Radioactive elements decay at a constant rate; at a given period of time, the 'half life', half the material will have decayed. It is possible to date rocks by considering the half life of the elements they contain.

heat. This new source of heat had obviously not been incorporated in Kelvin's calculations, which had been based on a simple cooling mechanism. This meant that the geological time-scale was much greater than Kelvin had supposed.

Radioactive elements emit radiation and change to a more stable 'daughter' element (see table below). The rate of this decay is constant and can be accurately measured. If the rate of decay is known and the proportions of the 'parent' radioactive element and the stable 'daughter' element in a rock are measured, it is possible to calculate how much time has elapsed since the rock was formed. In 1906 one of the first attempts at radiometric dating came up with a date of 2,000 million years for the rock being examined – far older than any date the geologists had postulated.

The calculations of the rate of decay are not as simple as might at first appear. The critical aspect of radioactive decay is the 'half life', which is the time it takes for half the radioactivity of the element to be emitted. This half life is a constant. Its use in dating naturally depends on the half life of the element being used. The gas Radon 220 has a half life of 52 seconds (that is, it only takes 52 seconds for half its radioactivity to be lost) and Strontium 90 of 27 years; both these half lives are far too low to be used for geological dating. Carbon 14 has a half life of 5,570 years and hence can be used for dating material between 1,000 and 70,000 years old. Potassium, with a half life of 1,300 million years, can only be used for material over 100,000 years. At the other end of the scale, Rubidium 87 has a half life of 47,000 million years and is only of use with material over 10 million years old.

The dating of rocks can be achieved by using different radioactive clocks and comparing the results. The rates of movement of tectonic plates can accurately date, for example, the opening of the Atlantic Ocean. Astronomical calculations involving the time light takes to travel from other galaxies can also confirm the geological time-span. The radiometric ages determined from Moon rock and meteorites give dates of 4,600 million years. As the Earth, the Moon and meteorites are all part of the solar system, it is assumed that their age is the same. There is in fact no evidence in support of Biblical chronology; it can be dismissed by simpler means, such as counting the annual bands of varved clays which total tens of thousands of years.

APPENDIX
LIVING FOSSILS

'Living fossil' is a contradiction in terms, since organisms are either living or fossil. The term, however, applies to plants and animals that are expected to have long been extinct yet still survive despite competition with more advanced forms. They are archaic forms that have been untouched by the process of evolution.

The most sensational discovery of a living fossil was made off the eastern coast of South Africa in 1938. At East London fishermen landed a large lobe-finned fish, which the curator of the local museum, Miss Latimer, recognized as a coelacanth, a primitive fish related to the ancestors of the land vertebrates, the last known fossil of which came from the Cretaceous. This specimen, of which only the skeleton and scales could be preserved, was named *Latimeria* in honour of Miss Latimer. Subsequently a deep water-dwelling population has been located off the Comoro Islands, north of Madagascar.

Another spectacular discovery was made in 1952. From a deep-sea trench in the Pacific off the coast of Mexico, a member of the most primitive group of molluscs, the Monoplacophora, was fished up. These primitive molluscs were thought to have become extinct during the Ordovician. This specimen had a limpet-like shell and provided the first evidence that the monoplacophorans showed signs of original segmentation which could perhaps link them with annelids. This animal was named *Neopilina*.

Other living fossils include the tuatara lizard *Sphenodon*, only known from islands off the coast of New Zealand; it seems to be the same creature as the primitive rhynchocephalian lizards of the Triassic, 200 million years ago. The first known brachiopod, *Lingula* from the Cambrian, has continued as far as can be judged quite unchanged throughout the fossil record. Similarly, the king crab *Limulus* has remained unchanged since the Carboniferous. Among the plants, the living *Gingko* is indistinguishable from its Jurassic forebears.

The reason certain organisms are termed living fossils is often a measure of their rarity. There are many others that are equally deserving of such an appellation, but are not so considered because of their familiarity. The harmless silverfish that inhabits houses is a primitive wingless insect

▲The horsetails *Equisetum* are a primitive group of plants that have survived unchanged from coal swamps of over 300 million years ago.

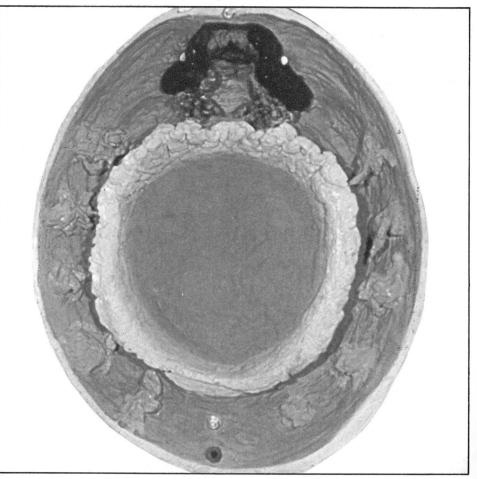

▲ *Neopilina* is a limpet-like mollusc belonging to the most primitive group of molluscs the Monoplacophora. They were thought to have died out 500 million years ago, until a live *Neopilina* was dredged up from the bottom of the Pacific in 1952.

▲ A living coelacanth. The coelacanths were common lobe-finned fish, which were thought to have died out 64 million years ago until a specimen was caught off the coast of South Africa in 1938.

▲ The North American opossum *Didelphis* is a small primitive marsupial which is still extending its range by becoming urbanized. This mammal has hardly changed since the Cretaceous 64 million years ago.

that has survived since the early Devonian. The cockroach, too, has remained unchanged since the Carboniferous. Among well-known plants, the horsetails are surviving members of the Carboniferous flora, the monkey-puzzle tree was established during the Triassic, and *Magnolia* represents one of the earliest true flowering plants from the Cretaceous. Among the vertebrates, the parasitic lamprey, the lungfishes of Africa, South America and Australia, tortoises and crocodiles, the American marsupial opossum and the insect-eating shrews are all survivors from the distant past that seem to have undergone little appreciable change.

It is often asked how, if evolution is the mechanism of survival, one can explain the existence of plants and animals from millions of years ago? There are several ways of tackling this problem. The most primitive forms of life, such as the bacteria that obtain their energy by breaking down organic molecules in the oxygen-free conditions at the bottom of stagnant ponds, flourished in a world devoid of oxygen. Today they still live where such conditions pertain. Similarly, the cyanobacteria or blue-green algae still inhabit hot volcanic springs or hypersaline waters. Just because more advanced types of organisms capable of living in oxygenated conditions have evolved, it does not mean there is no longer a place for organisms adapted to anoxic conditions. These still persist; in fact the evolution of higher forms of life can provide new environments for primitive organisms to exploit, for instance, the extensive bacterial flora which lives in the human body.

The essence of the process of evolution is that it is accumulative. It is for this reason that the diversity of animal and plant life on Earth is so vast. To take a simple example, the evolution of amphibians from fish did not mean that they took over all the environments inhabited by fish, nor when the reptiles evolved did it mean that the amphibian ecological niches were eliminated. But the most likely reason for the longevity of the primitive members of major groups is that they represent the basic stock that is unspecialized and therefore adaptable. Some descendants become adapted to certain narrow niches which in the course of events are subject to change or competition: specialization makes for success in the short term, but for longer-term survival adaptability is what matters. As the eminent zoologist Sir Gavin de Beer said, 'The paths of adaptive glory lead but to the grave of extinction.'

197

SUPPOSED SURVIVORS AND OTHER CURIOSITIES

Claims are made from time to time that prehistoric monsters have been sighted. One of the more notorious is the so-called Loch Ness Monster, which is said to be a surviving long-necked plesiosaur. The first 'proof' of the monster was a photograph taken by an anonymous surgeon in April 1934. However, one problem of observing objects in water is the difficulty of judging distance and scale. This famous 'plesiosaur' photograph is always assumed to be the head of an animal being lifted out of the water, but it could also be the tail of an animal going into the water and indeed zoologist Maurice Burton has suggested that it is an otter entering the loch.

The well-known photographs, first published in 1975, by Sir Peter Scott and Robert Rhines show a body, diamond-shaped fin and grotesque head. They are said to represent a long-necked reptile with oar-type paddles and a strangely gargoyle-like head, named *Nessiteras rhombopteryx* (an anagram of 'Monster hoax by Sir Peter S'). When the photographs are examined some strange similarities can be seen. The supposed body is reminiscent of the disintegrated hulk and prow of a Viking longboat, and the 2m-long right 'fin' seems more likely to have been a steering rudder positioned at the starboard stern (posterior right-hand side). Finally, the 'head' resembles the Öseberg head, an embellishment of Viking royal furniture.

It is obviously possible to interpret the evidence of the Loch Ness Monster differently when it is examined in the light of our knowledge of plesiosaurs. It is difficult to see how a descendant of inhabitants of warm seas could endure the cold waters of one of the most infertile lochs of Scotland. One should also remember that until 12,000 years ago Loch Ness did not exist. Before occupying Loch Ness, the surviving plesiosaurs would have had to live somewhere else for 64 million years in a world where their ecological niche was occupied by whales and sea-lions. Research on the swimming movements of both long- and short-necked plesiosaurs shows that plesiosaur limbs were hydrofoil-shaped, and not oar-shaped as has been suggested by Scott. It seems unlikely that the Loch Ness Monster is a plesiosaur, although obviously it would be a great find if it proved to be one.

Clearly preserved tracks of dinosaurs' feet have been found in the river-bed of the Paluxy River near Glen Rose in Texas, USA. Geologists are unanimous in saying that the river-bed must be dated to the end of the Mesozoic era in the Cretaceous – 140 million years ago. Footprints resembling those of man were found in the same stratum, close to the dinosaurs' footprints.

▲ A photograph taken by the London surgeon, R. K. Wilson on 1 April 1934, which purported to be the head and neck of the Loch Ness Monster; it is probably a bird or otter entering the water.

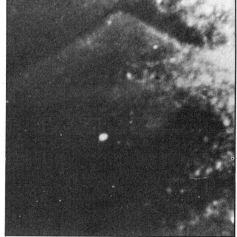

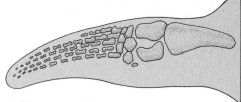

▲ Diagram of plesiosaur flipper based on skin impressions, compared with the oar-like flipper of *Nessiteras rhombopteryx*.

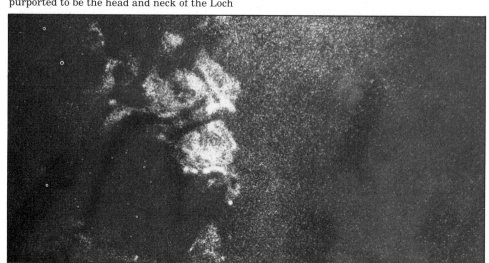

▲ Supposed head of the Loch Ness Monster, showing somewhat mammalian gargoyle-like appearance, photographed by Dr Robert Rhines in 1975.

▲ Sketch of the Viking Öseberg head which bears an uncanny resemblance to the supposed Loch Ness Monster head.

The author Erich von Daniken in his book *According to the Evidence* cites these footprints as evidence of the simultaneous appearance of man and dinosaur, 'About the time of the mass death of the dinosaurs there were very few mammals at first and according to the Darwinian theory there were certainly no men! Halstead: "No man ever saw a living dinosaur because the family of man did not exist in those days!" To err is human and there is nothing to be ashamed about in admitting an error. L. B. Halstead has erred here! Against this "dogma" inspired by Darwin, we can put facts, facts that are wilfully ignored because they are shaking the foundations of a sacrosanct theory. The facts in question are hard facts in the literal sense of the word.'

He has published photographs showing a track of oval prints some 450m in length 'walking' in the opposite direction, but alongside a giant bipedal dinosaur and taking identical strides. There seems little doubt that these are genuine prints of something. Dr Bob Slaughter, Director of the Shuler Museum of Palaeontology, Dallas, Texas, has noted that the heel-prints of bipedal, ornithopod, herbivorous dinosaurs appear rather like human prints but without toes and has further noted that since exposure, some of these prints have subsequently – and rather strangely – acquired toes. The authenticity of the prints with toes could be checked because original footprints are always preserved in a fossil algal mat which covers the overlying sediments on which they were made. To date such a testing has not been permitted by the Cretationists, who own the prints. Interestingly, in the British Museum (Natural History) London there is a trackway of *Iguanodon* footprints, on either side of which are two miniature human-like tracks. These are interpreted as the prints of the *Iguanodon* fore-limbs by English palaeontologist Dr David Norman. There is certainly no evidence available to state with any certainty that the footprints at Glen Rose establish the contemporaneity of man and dinosaur as stated in all the current Creationist literature and as put forward by von Daniken. The evidence seems to point to the trace fossils being those of dinosaurs.

A comparable case was put forward by the author Peter Kolosimo claiming that the skull of Rhodesian man, which has a neat round hole in the side of the skull, 'seems to demonstrate clearly the hole made by a bullet'. He proffered this as evidence of 'early cosmic encounters' 40,000 years ago. This particular interpretation ignores the fact that a bullet entering a skull would not make a neat hole, but would shatter it. The edge of the hole is smoothed and forms a slightly raised rim which is exactly how abscesses erode bone. Evidence of other abscesses was found around the teeth. As with the supposed human footprints, the evidence is not in dispute, but the interpretations are entirely misleading. Scientific experts should not simply ignore the challenge of these deceptive ideas, especially as they have such wide currency. Many people will not have the necessary information to refute erroneous views and the silence of the scientists might suggest that this is because they have no answer.

▼ Skull of Rhodesian man, a primitive *Homo sapiens*, showing a hole in the temporal bone due to bone resorption and similar erosion of bone around roots of teeth caused by abscesses.

▲ Dinosaur and elongated prints claimed to have been made by giant men from the Paluxy River, Texas, USA.

▶ Diagram of *Iguanodon* footprints from Swanage, Dorset, England, with small human-like prints made by the fore-limbs.

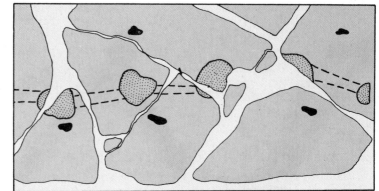

APPENDIX

CLASSIFICATION

The internationally recognized system of animal classification used today dates from the tenth edition of *Systema Natura* (1758) by the Swedish naturalist Carl Linnaeus (1707–78). He introduced the binomial Latin names of which the first is the genus or group name and the second identifies the individual species. For example, in the case of modern man or *Homo sapiens*, *Homo* is the genus 'Man' and *sapiens* the species. Just as the species are grouped in a genus, the genera are grouped in a family; for example *Homo* is in the family Hominidae. Families are grouped in orders (in our case, Primates), orders in classes (Mammalia), classes in a phylum (Chordata), and phyla in a kingdom (Animalia).

The phyla are the major divisions of the kingdom. One frequent challenge to evolutionists is to explain why there are no intermediates, or features in common, between phyla (although the living onychophoran *Peripatus*, has characteristics found in both arthropods and annelids). This is probably because intermediates never existed. If one takes the five most advanced animal phyla – the arthropods, annelids, molluscs, echinoderms and chordates – there is little evidence either in the fossil record or in the living fauna of links. However, the initial stages of the first three phyla – the arthropods, annelids and molluscs – are identical, for instance, the way in which the fertilized egg divides and the free-swimming larva (the trochophore) are the same; thereafter the developmental paths diverge. This evidence seems to indicate that these phyla have some real, albeit remote, connection in evolution. The echinoderms and chordates are also identical in their early stages, but then develop in different ways.

When it comes to dividing members of phyla into smaller groups, the way in which it is done is fundamentally arbitrary. The purpose behind the classification shapes the system. When Linnaeus devised his system, he structured it as a convenient *aide mémoire*, but he also thought he was uncovering God's scheme, believing that each species was individually created and immutable (although he did slightly change his mind later). If one takes the vertebrates, for example, they are divided into a series of major divisions known as classes: agnathans, fish,

amphibians, reptiles, birds and mammals. There is a qualitative difference in these divisions, in that each class is more advanced than the one before it, and is therefore known as a structural grade. The evolutionists first used Linnaeus' system to demonstrate lines of descent, the assumption being that from a single stock of, for instance, amphibian or

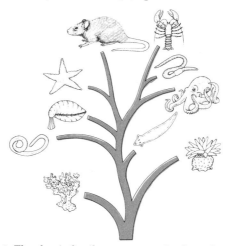

▲ The classic family tree suggesting how the phyla (major divisions) of the animal kingdom are related to one another. The phyla are shown to be distinct, but there is little evidence to support this scheme.

mammal, all subsequent descendants were amphibians or mammals. However, as more fossils were discovered, it was realized that in many instances more than one lineage frequently crossed these class boundaries. For instance, the ancestors of birds and mammals, although not directly related, were both reptiles. The rigid application of Linnaeus'

▲ Reptiles, Birds & Mammals can be divided horizontally into structural grades. An alternative method is to divide them vertically on the basis of genetic relationships or clades.

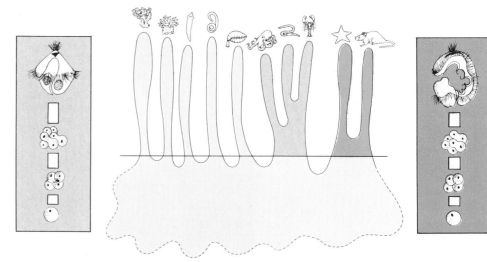

▲ An alternative family tree or fence of the phyla, which are shown as being derived along their independent pathways from the protist or unicellular stage. The diagram

shows that the molluscs, annelids and arthropods share the same early developmental stage, as do the echinoderms and chordates.

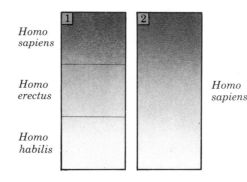

▲1 Evolutionary systematics; a single lineage can be divided into arbitrary sections.
2 Hennigian cladistics; a single unbranched lineage must be included within a single species.

▲ Confused cladistics; confusion is caused when species erected as arbitrary divisions of a continuum are treated as if they represented former evolutionary splitting events.

classification to evolving lineages could therefore be misleading.

It is possible to divide the evolutionary tree purely on the basis of ancestor-descendant relationships or lineages. The branch of the reptiles that gave rise to the birds and the living reptiles can be united with its descendants in a group called Sauropsida, while the mammal-like reptiles and the mammals can be grouped as Theropsida. One cannot then strictly separate the birds, for example, from the other Sauropsida as being more advanced, because this would be reverting to a structural grade interpretation of classification.

This system of classification is known as cladistics and was introduced in 1966 by W. Hennig in *Phylogenetic Systematics*. It disregarded gradual morphological change within a single lineage; new divisions were only to be named at the points of branching or speciation. Whereas the evolutionary systematists would divide an evolutionary lineage, for example, man, into arbitrary divisions, that is *Homo habilis, Homo erectus* and *Homo sapiens*, on the basis of major morphological changes such as in the brow, the hand and the brain size, under the new system there could be no gradual evolution directly from one species into another. Where a lineage could be adequately documented, the entire lineage had to be included in a single species, in this case *Homo sapiens*, with no other division. Where a form branched off from the parent species by becoming geographically isolated, like Darwin's Galapagos finches, or reproductively isolated, like the species *Australopithecus*, then at that point two new species come into existence by definition, even if there is no morphological change in the parent species. For instance, when *Australopithecus robustus* split off from another species, *Australopithecus africanus*, according to the evolutionists, there is one new species, *A. robustus*, and a continuation of the parent species, *A. africanus*. However, according to Hennig, there are two new species, *A. robustus,* and a species requiring a new name even though it is indistinguishable from *A. africanus*, which is deemed to have ended at the speciation. Hennig's method of classification disregards morphological features within a lineage if there are no splitting events. Both systems can be used to divide up the evolutionary tree, but it is important not to treat a species defined by one method as if it had been defined by another.

Unfortunately, untold confusion can be produced if species defined by evolutionary systematist methods are treated as if they were cladistic or Hennigian-type species. For example, in the case of Man, the three species *Homo habilis*, *Homo erectus* and *Homo sapiens* have been treated by cladists as the product of two separate speciation events. This interpretation means that no species can be seen as directly ancestral to any other, although anthropologists generally consider them to be in direct ancestor-descendant relationships. The cladist view leads to the astonishing assertion, warmly welcomed by the Creationists, that no ancestors of modern man are preserved in the fossil record. This is simply the result of the cladists not properly understanding their own methodology.

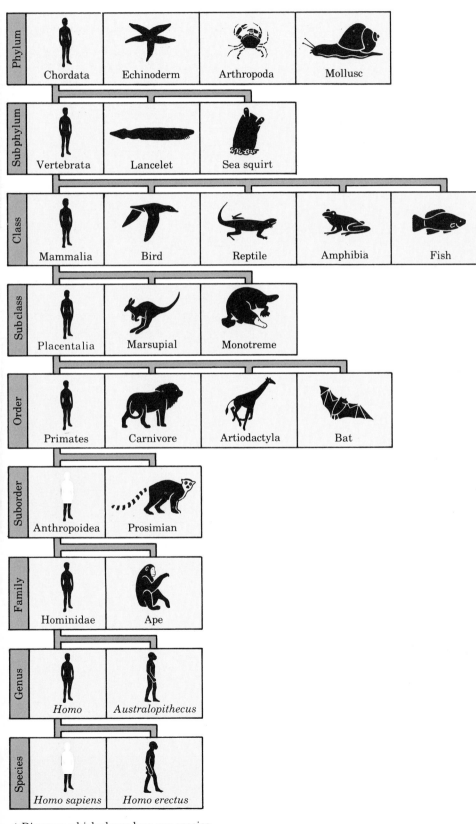

▲ Diagram which shows how our species *Homo sapiens* is classified according to the Linnaean system of classification.

GLOSSARY

Words in *italics* are defined elsewhere in the glossary.

Agnathan
A primitive jawless fish, such as the lamprey.

Alluvial
Pertaining to the sediments deposited by a river or stream.

Amphibian
A vertebrate animal, member of the class Amphibia, capable of living on land or in water, such as a frog or newt. Amphibians have a young aquatic gill-breathing stage and a lung-breathing adult stage.

Anaerobic
The absence of oxygen, and the ability to live without oxygen, such as anaerobic bacteria.

Anticline
Fold in a rock formation in the shape of an arch.

Arthropod
Invertebrate animal, member of the phylum Arthropoda, with an external skeleton, segmented body and jointed limbs, such as the insects and fossil trilobites.

Banded ironstone formation
A sequence of rocks made up of alternating iron-rich and iron-poor chemically produced sediments.

Biomass
The total weight or volume of organic material in a given area.

Bivalve
Invertebrate animal, member of the class Bivalvia, with a shell made up of two hinged halves.

Breccia
Rock composed of coarse angular fragments mixed with finer material.

Calcareous
Composed of or containing calcium carbonate ($CaCO_3$).

Caliche/calcrete
Carbonate enriched soil, in which the carbonate may cement the soil.

Cephalopod
Invertebrate animal, member of the class Cephalopoda, with well defined head and eyes and tentacles around the mouth. Fossil ammonites and the modern octopus are cephalopods.

Chitin
A horny substance forming the exoskeleton of *arthropods*.

Chloroplast
Small organelles found in the cells of plants, containing chlorophyll and involved in *photosynthesis*.

Chordate
Member of the phylum Chordata, which includes all those animals which at some stage in their development have an internal skeletal rod, the *notochord*.

Coelenterate
Invertebrate animal, member of the phylum Coelenterata, animals with a hollow body cavity and radial symmetry, including the sea-anemones, jellyfish and corals.

Conglomerate
Rock made up of coarse pebbles mixed with finer material.

Continental shelf
The submerged part of a continental land mass, covered to a depth of up to 200m by sea water.

Coprolite
Fossil excrement (droppings).

Cross-bedding
A series of large-scale inclined bedding surfaces, formed as the result of sediments avalanching down the front face of a wind or water produced sand-dune.

Cross-lamination
A series of small-scale inclined laminations formed in the same way as *cross-bedding*. *Cross-bedding* and cross-lamination can be seen in both ancient and recent sediments.

Cyanobacteria
Photosynthesizing *prokaryote* blue-green algae which are involved in the formation of *stromatolites*.

Death assemblage
An assemblage of *fossils* preserved away from their original habitat.

Decomposers
Organisms which obtain food by breaking down the remains and waste products of animals.

Deoxyribonucleic acid (DNA)
The large coiled molecule which makes up a single chromosome and in which is stored all the genetic information, or genetic code.

Diluvial/diluvialist
Pertaining to the Biblical Flood described in Genesis; person who believes certain geological phenomena can be explained by the Flood.

Echinoderm
Invertebrate animal, member of the phylum Echinodermata, with a skeleton of calcareous plates or spicules and a five-fold symmetry, such as the sea-urchins.

Ecology
The study of the interrelations of plants and animals and their physical environment.

Ecosystem
A system of ecological inter-relationships within a physically defined area – a garden pond and the world's oceans are both ecosystems.

Ectotherm
A cold-blooded animal which has to actively maintain a constant internal body temperature essentially by seeking shade when hot or sun when cool; *reptiles* are ectotherms.

Endotherm
A warm-blooded animal which can passively maintain a constant internal body temperature by internal metabolic processes; *mammals* are endotherms.

Eukaryote
Animal or plant with cell or cells that have a well defined nucleus containing its genetic material.

Exoskeleton
An external skeleton, protecting the soft part of the body, found mainly in *invertebrate* animals such as insects.

Food chain
A series of plants and animals that are linked together as food and feeder.

Food web
The often complex network of *food chains* found within an *ecosystem*.

Fossil
The preserved remains or traces of past life, more than 10,000 years old.

Fossil record
The evidence of past life which is available to us in the form of *fossils*.

Gastropod
An *invertebrate* animal, member of the class Gastropoda, usually with a single *calcareous* coiled shell as snails.

Igneous rock
Rock formed by the cooling and solidification of *magma*.

In-fauna
Animals living in soft sediments/soils.

Insectivore
Animal with a staple diet of insects.

Invertebrate
Animal without an internal vertebral column (backbone).

Isostasy
The tendency for the Earth's crust to maintain a balance between its uplifted and lowland areas.

Life assemblage
An assemblage of *fossils* containing species belonging to a single community preserved in the environment in which they lived.

Magma
Hot molten material formed beneath the Earth's surface.

Mammal
A *vertebrate* animal, member of the class Mammalia, which is warm-blooded and usually gives birth to live young which it suckles.

Meiosis
Process of cell division involved in the production of eggs and seeds during sexual reproduction, in which the parent cell divides to give four daughter cells with nuclei containing only half the number of chromosomes of the parent cell.

Metamorphic rock
Rock formed by the application of intense heat and/or pressure to existing *sedimentary* or *igneous* rocks.

Mitochondria
Small organelles found in animal cells concerned with the production of sugars for conversion to energy.

Mitosis
Process of cell division involved in

asexual reproduction in which the parent cell divides to give two daughter cells with nuclei containing the same number of chromosomes as the parent cell.

Neolithic
Or new Stone Age, part of the Pleistocene period during which agriculture became well established.

Notochord
A narrow, gelatinous rod which provides skeletal support in juvenile stages and in primitive chordates.

Ornithischian
Dinosaurs in which the pelvic bones have a three-pronged arrangement resembling that of birds.

Orogeny
A period of mountain building.

Outgassing
The expulsion of volatile gases which occurs when *magma* is cooling.

Outwash fan
A broad alluvial fan formed of sediments laid down by streams from a melting glacier.

Palaeolithic
Or old Stone Age, part of the Pleistocene when man first developed stone cultures.

Paramammals
Term applied to the group of mammal-like *reptiles* which ultimately gave rise to true *mammals*.

Permafrost
Layer of permanently frozen soil.

Photosynthesis
The production of sugars by plants using the Sun's energy, which takes place in the *chloroplasts*.

Phyletic gradualism
The concept that changes taking place in an evolutionary lineage, such as a new species, occur gradually.

Prokaryote
Organism with cells that lack a well defined nucleus; blue-green algae and bacteria are prokaryotes.

Punctuated equilibria
Concept that the *fossil record* shows long periods of stasis (equilibria) during which there is little or no change within a species, punctuated by shorter periods during which new species arise, and that this is the general pattern of evolution.

Reptile
A *vertebrate* animal, member of the class Reptilia, which is cold-blooded, has a scaly skin and lays eggs. Reptiles include lizards and snakes.

Saltation/saltationist
Leaps and jumps; a person who believes that evolution occurs thus.

Saurischian
Dinosaur in which the pelvic bones have a four-pronged arrangement, resembling that of *reptiles*.

Sedimentary rock
Rock formed from sediments derived from existing *metamorphic, igneous* and sedimentary rocks, and/or accumulations of organic debris.

Shoal
A submerged sand bank

Silica
An oxide of silicon (SiO_2), abundant in nature as quartz or silicates.

Stromatolite
A reef-like community built up by blue-green algae and bacteria.

Subfossil
The preserved remains or traces of past life, less than 10,000 years old.

Substrate
The upper surface of sediment or rock on which plants and animals live.

Symbiosis
An association between two organisms which is mutually beneficial to both.

Syncline
Fold in a rock formation in the form of a basin.

Tectonics
The study of the major structural features of the Earth's crust, or the term applied to the general structure of an area.

Till
An unstratified glacial deposit such as boulder clay.

Uniformitarianism
The concept that, 'The present is the key to the past', and that geological processes in the past have occurred at much the same rate as they do today.

Vertebrate
Animal with a vertebral column (backbone).

BIBLIOGRAPHY

*Advanced texts

GEOLOGY AND PLATE TECTONICS

* Allen, John R.L. *Physical Processes of Sedimentation*. Allen & Unwin, London, 1970 *Physical Geology*. Allen & Unwin, London, 1975 *Sedimentary Structures – Character and Physical Basis*. 2 vols. Elsevier, Amsterdam, 1982

* Anderton, R., Bridges, P.H., Leeder, M.R. and Sellwood, B.W. *A Dynamic Stratigraphy of the British Isles – a Study in Crustal Evolution*. Allen & Unwin, London, 1979

Anon. *The Story of the Earth*. Institute of Geological Sciences, London, 1972

Cocks, L.R.M. (ed.) *The Evolving Earth*. British Museum (Natural History), London and Cambridge University Press, Cambridge, 1981

Cox, C.B., Healey, I.N. and Moore, P.D. *Biogeography – an Ecological and Evolutionary Approach*. Blackwell, Oxford

Hallam, A. (ed.) *Planet Earth*. Rigby Ltd, Sydney, 1977; Elsevier, Oxford, UK, 1977

Imbrie, J. and Imbrie, K.P. *Ice Ages*. Macmillan, London and Basingstoke, 1979; Enslow Publications Inc., New Jersey, USA, 1979

Oxburgh, E.R. The Plain Man's Guide to Plate Tectonics. Geologists' Association, London, 1974: *Proceedings Geologists' Association* 85, 129–357

* Reading, H.G. (ed.) *Sedimentary Environments and Facies*. Blackwell Scientific Publications, Oxford, 1978

Tarling, D.H. and Tarling, M.P. *Continental Drift*. Doubleday & Co. Inc., New York, 1971; Penguin, London, 1972

Thackray, John. *The Age of the Earth*. Institute of Geological Sciences, London, 1980

HISTORY OF LIFE

Attenborough, David. *Life on Earth*. Collins, London, 1979; Little Brown & Co., Boston, 1978

* Brouwer, A. *General Palaeontology*. Oliver & Boyd, Edinburgh, 1967; University of Chicago Press, 1968

Halstead, L.B. *The Pattern of Vertebrate Evolution*. Oliver & Boyd, Edinburgh, 1969; W.H. Freeman, San Francisco, 1969

Halstead, Beverly and Middleton, Jennifer. *Bare Bones, an Exploration in Art and Science*. Oliver & Boyd, Edinburgh, 1972; Toronto University Press, 1973

* McKerrow, W.S. (ed.) *Ecology of Fossils*. Duckworth, London, 1978; MIT Press, Cambridge, Mass.

* Romer, A.S. *Vertebrate Paleontology*. University of Chicago Press, Chicago, 1966

* Romer, A.S. *Notes and Comments on Vertebrate Paleontology*. University of Chicago Press, Chicago, 1968

Steel, R. and Harvey, A.P. *The Encyclopaedia of Prehistoric Life*. Mitchell Beazley, London, 1979; McGraw-Hill Book Co., New York, 1979

* Watson, D.M.S. *Paleontology and Modern Biology*. Yale University Press, New Haven, 1951

* Young, J.Z. *The Life of Vertebrates*. Clarendon Press, Oxford, 1980

HISTORICAL

Bailey, Sir Edward. *Charles Lyell*. Thomas Nelson, London, 1962

Darwin, Charles and Huxley, T.H. *Autobiographies*. Oxford University Press, London, 1974

Darwin, Charles and Wallace, A. R. *Evolution by Natural Selection*. Cambridge University Press, Cambridge, 1958

Edwards, W.N. *The Early History of Palaeontology*. British Museum (Natural History), London, 1976

Gillespie, Charles C. *Genesis and Geology*. Harper & Row, New York, 1959

Greene, John C. *The Death of Adam. Evolution and Its Impact on Western Thought*. Iowa State University Press, Ames, 1959

Howard, Robert W. *The Dawnseekers, the First History of American Paleontology*. Harcourt, Brace Jovanovich, New York, 1975

Wendt, Herbert. *Before the Deluge*. Victor Gollancz, London, 1968

EVOLUTION

Cain, A.J. *Animal Species and Their Evolution*. Hutchinson, London, 1963

Darwin, Charles. *On the Origin of Species by means of Natural Selection, or the Preservation of favoured races in the struggle for life*. John Murray, London, 1859 (reprinted Penguin)

Engels, Friedrich. *Dialectics of Nature*. Foreign Languages Publishing House, Moscow, 1925 (1954 ed.); International Publishing Co., New York, 1940

* Gould, S.J. and Eldredge, N. Punctuated Equilibria: the Tempo and Mode of Evolution Reconsidered. *Paleobiology* Vol. 3, pp. 115–151, 1977

Hitching, F. *The Neck of the Giraffe or Where Darwin Went Wrong*. Pan Books, London, 1982

Ruse, Michael. *The Darwinian Revolution*. University of Chicago Press, Chicago, 1979

Ruse, Michael. *Darwinism Defended*. Addison-Wesley, Reading. Mass., 1982

* Scientific American *Evolution*. W.H. Freeman, San Francisco, 1978

* Stanley, S.M. *The New Evolutionary Timetable*. Basic Books, New York, 1981

FOSSIL COLLECTING

* Behrensmeyer, Anna K. and Hill, Andrew P. (eds.) *Fossils in the Making: Vertebrate Taphonomy and Paleoecology*. University of Chicago Press, Chicago, 1980

Halstead, L.B. The International Palaeontological Expedition to Sokoto State 1977–1978. *Nigerian Field Monograph*, no. 1, pp. 1–72, 1979

Kielan-Jaworowska, Zofia. *Hunting for Dinosaurs*. MIT Press, Cambridge, Mass., 1969

* Rixon, A.E. *Fossil Animal Remains: Their Preparation and Conservation*. Athlone Press, London, 1976

ORIGIN AND EVOLUTION OF LIFE

* Folsome, C.E. *The Origin of Life – a Warm Little Pond*. W.H. Freeman, San Francisco, 1979

* Halstead, L.B. *Vertebrate Hard Tissues*. Wykeham, London, 1974

* Margulis, Lynn. *Symbiosis in Cell Evolution*. W.H. Freeman, San Francisco, 1980

INVERTEBRATES

Clarkson, E.N.K. *Invertebrate Palaeontology and Evolution*. Allen & Unwin, London, 1979

PLANTS

Chaloner, W.G. and Macdonald, P. *Plants Invade the Land*. Royal Scottish Museum, Edinburgh, 1980

Thomas, Barry. *Evolution of Plants and Flowers*. Peter Lowe, London, 1981; St Martin's Press Inc., New York, 1981

FISH

* Miles, Roger S. *Palaeozoic Fishes*. Chapman & Hall, London, 1971

DINOSAURS

Charig, A.J. *A New Look at the Dinosaurs*. Heinemann, London, 1979; Smith Publishers, New York, 1979

Desmond, A.J. *The Hot-blooded Dinosaurs*. Blond & Briggs, London, 1975

Halstead, L.B. *The Evolution and Ecology of the Dinosaurs*. Peter Lowe, London, 1975

Halstead, L.B. and Halstead, Jenny. *Dinosaurs*. Blandford, Poole, UK, 1981

Russell, D.A. *A Vanished World – the Dinosaurs of Western Canada*. National Museums of Canada, Edmonton, Alberta, 1977

* Thomas, R.D.K. and Olsen, E.C. *A Cold Look at Warm-blooded Dinosaurs*. Westview Press, Boulder, Colorado, 1980

MAMMALS

Halstead, L. B. *The Evolution of the Mammals*. Peter Lowe, London, 1978

* Lillegraven, Jason A., Kielan-Jaworowska, Zofia, and Clemens, William A. (eds.) *Mesozoic Mammals*. University of California Press, Berkeley and Los Angeles, 1979

* Scott, W.B. *A History of Land Mammals in the Western Hemisphere*. Hafner Publishing Co., New York, 1937

Simpson, G.G. *Horses*. Oxford University Press, New York, 1951

Splendid Isolation – the Curious History of South American Mammals. Yale University Press, New Haven, 1980

* Stuart, A.J. *Pleistocene Vertebrates in the British Isles*. Longman, London, 1982

MAN

* Bishop, W.W. (ed.) *Geological Background to Fossil Man*. Geological Society, London, 1978

Cole, Sonia. *The Neolithic Revolution*. British Museum (Natural History), London, 1961

* Cronin, J.E., Boaz, N.T., Stringer, C.B. and Rak, Y. Tempo and Mode in Hominid Evolution. *Nature*, Vol. 292, pp. 113–122, 1981

Hamblin, D.J. *The First Cities*. Time-Life International, Amsterdam, 1972

Howell, F. Clark. *Early Man*. Time-Life, Amsterdam, 1966

Leonard, J.N. *The First Farmers*. Time-Life, Amsterdam, 1972

Oakley, K.P. *Man the Tool-maker*. British Museum (Natural History), London, 1958; University of Chicago Press, 1976

Reader, John. *Missing Links, the Hunt for Earliest Man*. Collins, London, 1981; Little Brown & Co., Boston, 1981

Watson, William. *Flint Implements, an Account of Stone Age Techniques and Culture*. British Museum, London, 1968

Wood, Bernard. *The Evolution of Early Man*. Peter Lowe, London, 1976

Wood, P., Vaczek, L. and Hamblin D. J. *Life Before Man*. Time-Life, Amsterdam, 1972

* Young, J.Z., Jope, E.M. and Oakley, K.P. (eds.) The Emergence of Man. *Philosophical Transactions of the Royal Society of London*, Biological Sciences, Vol. 292, pp. 1–216, 1981

Selected research monographs, papers and articles by L.B. Halstead, on which both data and theoretical conclusions presented in this book have been based.

* Psammosteiformes (Agnatha), a review with descriptions of new material from the Lower Devonian of Poland. I. General Part. *Palaeontolgia Polonica*, vol. 13, pp. vii & 1–135, 1965

* Psammosteiformes (Agnatha), a review with descriptions of new material from the Lower Devonian of Poland. II. Systematic Part. *Palaeontologia Polonica*, vol. 15, pp. ix & 1–168, 1966

Major Faunal Provinces in the Old Red Sandstone of the Northern Hemisphere. *Int Symp. Devonian System*, Calgary, vol. 2, pp. 1231–1238, 1967

The Mitochondrion and the Origin of Bone. *Calc. Tiss. Res.*, vol. 3, 103–105, 1969

* Calcified Tissues in the Earliest Vertebrates. *Calc. Tiss. Res.*, vol. 3, pp. 107–124, 1969

The Presence of a Spiracle in the Heterostraci (Agnatha). *Zool. J. Linn. Soc.*, vol. 50, pp. 195–197, 1971

with Nicoll, P.G. Fossilized Caves of Mendip. *Stud. Speleol.*, vol 2, pp. 93–102, 1971

* The Heterostracan Fishes. *Biol. Rev.*, vol. 48, pp. 279–322, 1973

* with Turner, Susan. Silurian and Devonian Ostracoderms. pp. 67–79, *Atlas of Palaeobiogeography*, ed. A. Hallam, Elsevier, Amsterdam, 1973

with Goriup, P.D. and Middleton J. A. The Loch Ness Monster. *Nature*, vol. 259, pp. 75–76, 1976

New Light on the Piltdown Hoax? *Nature*, vol. 276, pp. 11–13, 1978

The Piltdown Hoax: Cui Bono? *Nature*, vol. 277, p. 596, 1979

* with Liu, Y.-H. and P'an, K. Agnathans from the Devonian of China. *Nature*, vol. 282, pp. 831–833, 1979

* Internal Anatomy of the Polybranchiaspids (Agnatha, Galeaspida). *Nature*, vol. 282, pp. 833–836, 1979

Popper: Good Philosophy, Bad Sciences? *New Scientist*, vol 87, pp. 215–217, 1980

Museum of Errors. *Nature*, vol 288, p. 208, 1980

Karl Popper, Palaeontology and Evolution. *Inter. Symp. Concpt. Meth. Paleo.* Barcelona, pp. 9–18, 1981

* Evolutionary Trends and the Phylogeny of the Agnatha. In *Problems of Phylogenetic Reconstruction*, Joysey, K.A. and Friday, A.E. (eds), pp. 159–196. Systematics Association Special Volume no. 21, Academic Press, London and New York, 1982

INDEX

PRINTED IN BELGIUM BY

proost
INTERNATIONAL BOOK PRODUCTION